Biomaterials, artificial organs and tissue engineering

Related titles:

Corrosion in reinforced concrete structures
(ISBN-13: 978-1-85573-768-6; ISBN-10: 1-85573-768-X)
In this authoritative new book the fundamental aspects of corrosion in concrete are analysed in detail. An overview of current monitoring techniques, together with a discussion of practical applications and current numerical methods that simulate the corrosion process, provides the civil and structural engineer with an invaluable guide to best practice when it comes to design aimed at minimising the effects of corrosion in concrete. The corrosion protective properties of concrete and modified cements are also discussed. The most frequently used stainless steels are examined together with an analysis of their reinforcement properties. Special attention is given to their handling and welding requirements, and the economics of their use. A comprehensive overview of surface treatments and corrosion inhibitors is presented alongside their practical applications and detailed coverage of electrochemical protection and maintenance techniques is provided.

Durability of engineering structures: design, repair and maintenance
(ISBN-13: 978-1-85573-695-5; ISBN-10: 1-85573-695-0)
Structures often deteriorate because little attention is given to them during the design stage. Most standards for structural design do not cover design for service life. Designing for durability is often left to the structural designer or architect who does not have the necessary skills, and the result is all too often failure, incurring high maintenance and repair costs. Knowledge of the long-term behaviour of materials, building components and structures is the basis for avoiding these costs. This book is aimed at degree students in civil engineering, engineers, architects, contractors, plant managers, maintenance managers and inspection engineers.

Advanced polymer composites for structural applications in construction
(ISBN-13: 978-1-85573-736-5; ISBN-10: 1-85573-736-1)
The proceedings of ACIC 2004 focus on the application and further exploitation of advanced composites in construction. With over 100 delegates, the conference, held at the University of Surrey from 20th to 22nd April 2004, brought together practising engineers, asset managers, researchers and representatives of regulatory bodies to promote the active exchange of scientific and technical information on the rapidly changing scene of advanced composites in construction.

Details of these and other Woodhead Publishing materials books and journals, as well as materials books from Maney Publishing, can be obtained by:

- visiting www.woodheadpublishing.com
- contacting Customer Services (e-mail: sales@woodhead-publishing.com; fax: +44 (0) 1223 893694; tel.: +44 (0) 1223 891358 ext. 30; address: Woodhead Publishing Limited, Abington Hall, Abington, Cambridge CB1 6AH, England).

If you would like to receive information on forthcoming titles, please send your address details to: Francis Dodds (address, tel. and fax as above; email: francisd@woodhead-publishing.com). Please confirm which subject areas you are interested in.

Maney currently publishes 16 peer-reviewed materials science and engineering journals. For further information visit www.maney.co.uk/journals.

Biomaterials, artificial organs and tissue engineering

Edited by
Larry L. Hench and Julian R. Jones

Woodhead Publishing and Maney Publishing
on behalf of
The Institute of Materials, Minerals & Mining

CRC Press
Boca Raton Boston New York Washington, DC

WOODHEAD PUBLISHING LIMITED
Cambridge England

Woodhead Publishing Limited and Maney Publishing Limited on behalf of
The Institute of Materials, Minerals & Mining

Published by Woodhead Publishing Limited, Abington Hall, Abington,
Cambridge CB1 6AH, England
www.woodheadpublishing.com

Published in North America by CRC Press LLC, 6000 Broken Sound Parkway, NW,
Suite 300, Boca Raton, FL 33487, USA

First published 2005, Woodhead Publishing Limited and CRC Press LLC
© Woodhead Publishing Limited, 2005
The authors have asserted their moral rights.

British Library Cataloguing in Publication Data
A catalogue record for this book is available from the British Library.

Library of Congress Cataloging in Publication Data
A catalog record for this book is available from the Library of Congress.

Woodhead Publishing Limited ISBN-13: 978-1-85573-737-2 (book)
Woodhead Publishing Limited ISBN-10: 1-85573-737-X (book)
Woodhead Publishing Limited ISBN-13: 978-1-84569-086-1 (e-book)
Woodhead Publishing Limited ISBN-10: 1-84569-086-9 (e-book)
CRC Press ISBN-10: 0-8493-2577-3
CRC Press order number: WP2577

The publishers' policy is to use permanent paper from mills that operate a sustainable
forestry policy, and which has been manufactured from pulp which is processed using
acid-free and elementary chlorine-free practices. Furthermore, the publishers ensure
that the text paper and cover board used have met acceptable environmental
accreditation standards.

Typeset by Replika Press Pvt Ltd, India

Contents

7 Inflammation and wound healing 71

JUNE WILSON HENCH, Imperial College London, UK

Part II Clinical needs and concepts of repair

8 The skeletal system 79

LARRY L. HENCH, Imperial College London, UK

9 The cardiovascular system 90

M. JOHN LEVER, Imperial College London, UK

17   Cardiovascular assist systems   179
M. JOHN LEVER, Imperial College London, UK

Part IV Tissue engineering

18   Introduction to tissue engineering   193
LEE D.K. BUTTERY, University of Nottingham, UK and
ANNE E. BISHOP, Imperial College London, UK

19   Scaffolds for tissue engineering   201
JULIAN R. JONES, Imperial College London, UK

1

Introduction

L A R R Y   L .   H E N C H
Imperial College London, UK

1.1   Book and CD module

This book takes a novel approach to the teaching of the multi-disciplinary subjects of biomaterials, artificial organs and tissue engineering. These three areas of the broad field of biomedical engineering are included together in this volume because of their scientific, technological and clinical interdependence. For example, biomaterials are critical components in artificial organs and as scaffolds in tissue engineering. Understanding the interfacial interaction of living cells with man-made materials is fundamental to all three fields. Rapid advances in cell and molecular biology have an impact on the development of new biomaterials and their use in prostheses, artificial organs and tissue-engineered constructs. Keeping pace with rapid developments in these fields is a challenge.

The goal of this book is to meet this challenge in three ways:

1. The printed volume consists of 25 chapters organised into five parts. The five parts provide an introduction to and overview of living and man-made biomaterials; the various systems of the human body and their clinical needs for repair; various types of implants, devices and prostheses used to repair the body; the principles underlying alternative types of artificial organs; the scientific basis and applications of tissue engineering; followed by a discussion of socio-economic and ethical issues that influence the repair and replacement of parts of the human body. Each chapter corresponds to a 1- or 2-hour lecture in a one semester course at an advanced undergraduate level.

2. Each of the 25 chapters in the book is supplemented by an extensively illustrated lecture on CD-ROM. The CD-ROM is attached to the inside back cover of this book.

3. The CD-ROM also contains self-study questions for each chapter. These questions are designed to test your knowledge of each chapter and consolidate what you have learnt. Answers are provided separately so that you can review what you have learnt from each chapter.

The combined CD-ROM supplementary lectures, self-study questions and answers are referred to hereafter as the CD module.

1.2   Aims and objectives

The aims of the book and CD module are to:

1. Provide an understanding of the needs, uses and limitations of materials and devices to repair, replace or augment the living tissues and organs of the human body.
2. Establish the biomechanical principles and biological factors involved in achieving long-term stability of replacement parts for the body.
3. Analyse the principles and applications of engineering of tissues to replace body parts.
4. Examine the industrial, governmental and ethical issues involved in use of artificial materials and devices in humans.
5. Enhance the reader's awareness of the complex multi-disciplinary nature of the fields of biomedical materials, biomechanics, total joint replacements, artificial organs and tissue engineering, and the importance and methods of communication with a broad range of people including clinicians, patients, industrial and governmental personnel.

Core objectives of the textbook and CD module

After studying this book and CD module the student will be able to:

- Identify various components of the human body, describe their function and explain the effects of ageing on the structure and function of various groups of tissues and organs.
- Describe the major classes of biomedical implant materials, their means of fixation, their stability and advantages, and their disadvantages when used as implants, devices and in artificial organs.
- Explain the types of failure of implants, devices and prostheses in various clinical applications and the reasons for failure.
- Describe the physiological principles involved in the replacement of various parts of the body with artificial organs, transplants or tissue-engineered constructs and the clinical compromises involved.
- Interpret lifetime survivability data of medical devices, and relate clinical success to biomaterial characterisation, device design and clinical variables.
- Discuss the clinical and socio-economic factors in the use of implants, transplants, artificial organs and tissue-engineered constructs including industrial, governmental, regulatory and ethical issues.
- Defend the relative merits of replacing a body part with a tissue-engineered

construct, discuss the principles involved in growing body parts *in vitro* and describe the physiological and clinical limitations involved.

- Research the literature for new developments in replacement of tissues and organs.
- Communicate alternative means to repair or replacement of parts of the body to healthcare professionals, patients, business and government personnel.

Specific objectives

There are five parts to the book and CD module. The following list documents what the student will be able to do by the end of each part.

Part I: Introduction to materials (living and non-living)

- Describe the microstructure of metals, ceramics, polymers and composite materials.
- Relate the mechanical strength and fatigue of various materials to their microstructure.
- Compare the processing routes for manufacture of various materials and explain how processing affects microstructure, properties and reliability of the materials.
- Describe the parts of a cell and their importance in cell function, proliferation and differentiation.
- Identify the four types of tissues and describe their function.
- Define the concept of biocompatibility and describe how it is tested.
- Explain the mechanism and relevance of inflammatory changes to tissues in contact with biomaterials and their importance in wound healing and long-term prognosis for implants and artificial organs.

Part II: Clinical needs and concepts of repair

- Define and compare the differences between implants and transplants and their relative advantages and disadvantages.
- Classify the types of tissues and organs in the body and their function in relation to replacement by artificial materials.
- Describe the effects of ageing on the structure and properties of the skeletal, cardiovascular, sensory and other systems.
- Identify the various categories of implant materials and devices, and the means of their fixation to various physiological systems.

Part III: Applications

- Identify the bones, tendons and ligaments of the human skeletal system and describe their mechanical properties as a function of ultrastructure, microstructure and macrostructure.

- Describe the repair mechanisms of the skeletal tissues as a function of mode of failure and age of patient.
- Select the types of fixation and joint replacement device required for various types of repair of the skeletal system and explain the advantages and disadvantages of use of each type of device related to the clinical problem addressed.
- Compare the physical properties of metallic alloys, medical polymers and bioceramics used for orthopaedic fracture fixation and total joint replacement.
- Analyse the sources of failure of orthopaedic prostheses.
- Describe the effects of elastic modulus on stress shielding of bone.
- Compare the relative merits of different total hip, total knee, shoulder and spine systems on long-term survivability by use of Kaplan–Meier survivability curves.
- Define bioactivity and compare bioactive fixation versus cement or morphological fixation of prostheses.
- Discuss polymeric materials that do not degrade in the physiological environment and their application in medical devices.
- Describe polymeric materials that undergo controlled rates of resorbtion in the body, such as those used in sutures, and the mechanisms of resorbtion and the effect of time and load on resorbtion rates.
- Describe the mechanisms of surface reactions that are involved in blood clotting and the surface characteristics of polymers that are haemocompatible and the blood contacting devices made from such materials.
- Understand how mechanical devices can replace the function of organ systems in the body, including the heart, blood vessels, lungs, kidneys and sensory organs.
- Describe the engineering design principles for effective function of artificial organs, including their performance in terms of their material properties, exchange processes, maintenance of appropriate blood flow and compatibility with the overall function in the body.
- Be capable of describing to healthcare professionals and patients the advantages, disadvantages and limitations of artificial organs.

Part IV: Tissue engineering

- Discuss for any given tissue engineering application the importance of evaluating sources of cells to promote repair, including somatic stem cells, embryonic stem cells and 'mature' cells from the tissue of interest.
- Discuss the basic principles of cell culture and characterise the various cell types using methods such as immunocytochemistry, scanning electron

and confocal microscopy to identify cell-type specific markers and genes.

- Analyse the processes of cell proliferation, cell differentiation, extracellular matrix synthesis and cell death, and discuss the relationship between these events.
- Discuss the importance of being able to exercise control over mechanisms of cell proliferation, differentiation and death in developing a tissue repair strategy and identify methods to achieve this by monitoring cell cytokines.
- Understand the importance of translating data derived from monolayer cell cultures into clinically usable cell/tissue types.
- Discuss interactions between cells and scaffolds and understand how scaffold composition and architecture can influence cell attachment, growth and differentiation.
- Discuss specific examples where tissue engineering strategies are being developed first to promote repair and regeneration with emphasis on bone, cartilage and skin and second to understand the need to evaluate mechanical strength and tissue integration.

Part V: Societal, regulatory and ethical issues

- Outline the steps required to achieve governmental regulatory approval for human use of a medical device.
- Chart the five pathways involved in technology transfer of a new biomedical material or device from the laboratory into production and assess the cost and time lines involved in each pathway.
- List the economics, social and regulatory barriers to bringing a new biomedical material, device or tissue engineering construct into the marketplace or clinic.
- Summarise various legal issues that limit the use of implantable devices, artificial organs and tissue-engineered constructs.
- Understand the concept, application and limitation of the three general principles of ethical theory.
- Debate the ethical issues involved in use and misuse of implants, devices, artificial organs and tissue-engineered constructs using principles of ethical theory and moral philosophy.
- Analyse various clinical cases requiring use of implants, transplants or tissue-engineered constructs in terms of scientific, economic and ethical principles.

1.3   Integrated use of book and CD module

The book and CD module are intended to be used together as a learning aid. Each book chapter is supplemented by a lecture, self-study questions and answers. The relationship between a book chapter and the equivalent section on the CD module is as follows:

Book	CD module
Chapter 2 Metals	Chapter 2 Metals
2.1 Introduction	PowerPoint® lecture
2.2 The metallic bond	Supplementary lecture text and
2.3 Microstructure	figures
2.4 Mechanical properties	Self-study questions
2.5 Fatigue properties	Self-study questions for Chapter 2
2.6 Hardness and wear	Answers to self-study questions
2.7 Shape memory effect and	Answers to self-study questions for
superelasticity	Chapter 2
2.8 Corrosion	
2.9 Effects of processing on	
structure, properties and	
reliability	
2.10 Clinical requirements	
2.11 Summary	
2.12 Reading list	

In each book chapter you will find cross-references to the equivalent lecture in the CD module. Some figures in book chapters also appear in the equivalent PowerPoint® lecture for the chapter. In these cases, you will see both the main figure number and (in brackets) its number in the PowerPoint® lecture. These cross-references are provided to help you to relate the relevant PowerPoint® lecture to the book chapter.

1.4   Using the CD-ROM

1. Insert the CD-ROM into the CD-ROM drive.
2. The CD-ROM should auto-run. If the CD-ROM does not auto-run, open Microsoft Internet Explorer® on your computer and open the file *index.html*.
3. Follow the instructions on the homepage.

Please note that the lectures are in Microsoft PowerPoint® and the self-study questions and answers are in PDF format.

1.5   Recommendations for self-study

The reader should use the book and CD module in the following way:

1. It is suggested that the reader start with the book, which provides an overview of the important topics in the subject.
2. For each chapter, study the relevant PowerPoint® lecture.
3. Read through and answer the self-study questions.
4. Check your answers to the self-study questions.
5. If you calculate your score to be less than 80%, it is recommended that you go back over the relevant material.

If you follow these steps, you should have understood the subject matter at sufficient depth to read the current literature in the field and discuss, and perhaps even debate, the topic with experts.

1.6   Overview

Since the 1960s the use of biomedical materials to repair, replace or augment diseased, damaged or worn-out body parts has increased enormously. More than 40 different materials are currently used to replace more than 40 different parts of the body. The worldwide number of implanted devices is in the range of 4 000 000 to 5 000 000 per year. Many millions of patients have enjoyed alleviation of pain, increased function and an improved quality of life due to the success of biomedical materials and the implants and devices made from them. Artificial organs are used routinely to prolong life and maintain quality of life for many thousands of patients when organ transplantation is not possible. Advances in biology and biomedical materials now make it possible to grow living constructs outside the body and to use them as engineered tissues for repair or replacement of diseased or damaged tissues.

The objective of this book and CD module is to provide a short summary of the use of man-made materials as medical implants in various clinical applications, including artificial organs and tissue engineering. The book is designed to be used in a one-term course at a third- or fourth-year undergraduate or first-year postgraduate level. It is assumed that many students may have only a minimal level of background in either materials or biological sciences. Consequently Part I (Chapters 2–5) summarises the metals, ceramics, polymers and composites used in medical applications. Chapter 6 reviews the cells and types of tissues and their hierarchical structures, which influence the interfacial behaviour of materials in the body. Chapter 7 summarises the mechanisms of wound healing and inflammatory response of tissues to implants. If you already understand these topics, you should proceed directly to Part II.

Part II emphasises clinical needs and the concepts of repair and replacement of bones and joints, the cardiovascular system and other physiological systems involving soft connective tissues. Because of the importance of biomechanics and the quality of skeletal tissues to the long-term success of orthopaedic implants, Chapter 8 reviews the structure and biomechanical properties of bone, cartilage, ligaments and tendons and their physiological mechanisms of repair. Chapter 9 describes the structure and function of the heart and arterial system and their modes of failure. Chapters 10 and 11 discuss the use of polymers and hydrogels as implant materials.

Part III consists of six chapters that describe the materials and methods used to repair bones, tendons and ligaments (Chapter 12), the replacement of joints (Chapter 13), the engineering principles of artificial organs (Chapter 14), mass transport processes in artificial organs (Chapter 15), artificial exchange systems for organ replacement (Chapter 16) and cardiovascular assist systems (Chapter 17).

Part IV provides an introduction to the newly emerging field of tissue engineering (Chapter 18), the types of scaffolds that can be used as templates for tissue growth (Chapter 19), the methods used to grow cells and tissues outside the body (Chapter 20), and methods for analysing and optimising cell behaviour to obtain tissue growth (Chapter 21). Chapter 22 describes the main applications of tissue engineering techniques for the regeneration of body parts and the clinical successes.

The final part of the book and CD module discusses many of the societal issues involved in the development, commercialisation and clinical use of implants and transplants. Assurance of reproducibility of properties and long-term clinical performance requires the establishment of a materials characterisation and quality assurance programme. All medical devices must comply with international standards and satisfy regulatory requirements, discussed in Chapter 23. The many steps and potential barriers to transfer new biomedical materials technology from a laboratory discovery into commercial clinical use are described in Chapter 24. The ethical principles that govern the use of biomedical devices, implants and transplants are discussed in Chapter 25 with examples of ethical concerns.

1.7   Are transplants the solution to spare parts?

One option for replacing diseased or damaged body parts is to use already living tissue often in the form of an organ, as the 'spare part'. A heart–lung transplant is an example.

A '*transplant*' is a tissue or an organ moved from one body, or body part, to another. Coronary bypasses using the patient's own sapphanous veins or spinal fusion using bone from the iliac crest are examples.

Alternative types of transplants

'*Autograft*' (transplant of tissue or organ from one part of a person to another location in the same person). Autografts are often considered the 'gold standard' in surgery. However, there is limited availability and autografts may deteriorate with time due to differences in their physiological loads and the biological environment in the transplant site.

'*Homograft*' (transplant of tissue or an organ from one human to another). Genetic differences give rise to rejection by the immune system and availability are the primary factors limiting use of homografts.

'*Heterograft or Xenograft*' (transplant of tissue or an organ from another species to human). There are increased genetic concerns, potentially circumvented by genetic engineering of the animal creating a 'transgenic species', that limit the use of heterografts. At present technical and ethical issues limit the use of heterografts to non-living tissue replacements, e.g. porcine heart valves and demineralised bovine bone as a bone graft substitute.

1.8   Implants and prostheses as spare parts

Because of limited supply, immune rejection and other concerns regarding use of transplants, most of the need for human spare parts is provided by implants; nearly 5 000 000 per year.

Implants are defined as 'A man-made material or device inserted or embedded surgically in the body'. Because implants are man-made and do not contain foreign proteins, they therefore do not elicit an immune response from the host.

'*Prosthesis*': A man-made device used inside the body to replace, repair or augment a diseased, damaged or missing part. (Implants and prostheses are often used as synonyms).

Examples of the large number of implants or prostheses implanted annually in the USA include: heart pacemakers (200 000), heart valves (40 000), intraocular lenses, IOL (1 000 000) and middle ear aeration (myringotomy) tubes (CD Fig. 1.5). The number of implants used annually in the USA for the repair of bones and joints is equally large. Examples are: joint replacement (500 000), temporary fixation devices (1 000 000) and spinal surgery (400 000) (CD Fig. 1.5). The annual number of implants in Europe is approximately the same as in the USA.

Several factors have led to a rapid increase in the use of implants and prostheses:

1. A population where the number of persons older than 50 years of age is rapidly increasing. This is due to an increase in average age per person combined with a worldwide growth in population (CD Fig. 1.4).
2. A progressive deterioration in the quality of connective tissues with age (CD Fig. 1.6–1.10). CD Fig. 1.6 illustrates the large changes in bone that occur from the age of 30 onwards. Bone loss is especially severe for post-menopausal women due to hormonal factors; as much as 50% loss

of bone mass may occur by the age of 70. This condition is termed 'osteoporosis' and is the cause of fracture of long bones, hips and collapse of vertebrae in millions of women.

3. An increase in the confidence of surgeons that implants will provide an improved quality of life for patients. This is partly due to the fact that there is a high survivability for many prostheses. For example, orthopaedic devices implanted with the use of polymethylmethacrylate (PMMA) 'bone cement' can be partially loaded by the patient within a day after surgery (CD Fig. 1.23, 1.24). An ambulatory patient recovers from the operation more quickly and the bearing of weight stimulates the bone to repair. Use of orthopaedic implants makes it possible to alleviate the chronic pain associated with many degenerative diseases of joints. The large number of intraocular lenses and their long survivability greatly enhance the vision of millions of elderly patients and leads to an impression that all replacement parts will be equally successful (CD Fig. 1.14, 1.15).

4. An increase in the confidence of patients that prostheses are acceptable, in fact are desirable, solutions to chronic pain.

5. Improvements in surgical skills, equipment and facilities have minimised most of the complications associated with major surgery, such as infection. Consequently, there is good long-term success of life-saving devices such as heart-pacers (CD Fig. 1.16) and prosthetic heart valves (CD Fig. 1.17).

6. Improvements in materials, designs, and post-surgical regimens have greatly increased lifetimes of devices during the last 20 years.

7. International standards and governmental regulations have led to high reliability of prostheses performance.

1.9   The limitations of implants

Implants are limited in function and survivability because they are not living. Natural living tissues have a genetically programmed capacity for self-repair. Consequently, natural tissues can adapt to their physiological environment. No man-made material is capable of self-repair or adaptation. The interface between prosthesis and its host tissue is especially susceptible to stress. Mismatch in either biochemical or biomechanical factors can lead to interfacial deterioration and eventual failure.

Thus, the emphasis in Parts II and III of this book is on the mechanisms of interaction of implanted materials with their host tissues. Changes in tissue occur at all levels as they repair at the surgical site and adapt to the implant. Biomechanical and biochemical alterations in the tissues are a function of the type of interface created, which is discussed in the first chapter of the CD module. The interfacial response to an implant also depends on the quality of the host tissue and the inflammatory response to the implant, described in Chapter 7. Because of these limitations, there is great interest in the potential for regenerating tissues by enhancing the body's own cellular repair mechanisms. This new technology, called tissue engineering and regenerative medicine, is described in Part IV.

All methods of repair or replacement of parts of the body currently in use have limitations. Thus, there is need to understand the reason for the limitations and the consequences when failure occurs. These technical and clinical factors are addressed in Parts II–IV. It is also necessary to understand the economic and regulatory factors that influence the use of implants and transplants. These are discussed in Part V. Because of the finite survivability of implants and transplants, there are ethical issues that must be considered during the development, testing and clinical use of all methods of repair or replacement of human body parts. Chapter 25 reviews the ethical principles that govern the decision-making process and the safeguards to protect the patient and the surgeon.

Part I

Introduction to materials
(living and non-living)

2

Metals

E. JANE MINAY and
ALDO R. BOCCACCINI
Imperial College London, UK

2.1   Introduction

Metallic materials are inorganic substances, usually combinations of metallic elements (such as iron, titanium, aluminium, gold), which may also contain small amounts of non-metallic elements (such as carbon, nitrogen and oxygen). Metals are rarely used as a pure element but are mixed with other elements to form an alloy. This is usually necessary to obtain the required properties of the material. For example, iron is alloyed with chromium to form stainless steel, and zinc with copper to form brass.

Three factors influence the choice of metals and alloys as biomaterials: (i) physical and mechanical properties, (ii) degradation of the material and (iii) biocompatibility. The microstructure of the metallic material, its properties and the processing routes used are all very strongly interrelated. This chapter introduces the reader to some general metallurgical principles that explain some of the mechanisms behind these factors. These can be applied to the wide variety of metallic systems but there will be a specific emphasis on the metals, properties and processes relevant to biomedical applications.

In general, metallic materials are used for orthopaedic applications, where their high strength is essential, in internal electrical devices, in orthodontics and in artificial organs. At present, the most important strong 'biometals' include stainless steels, cobalt alloys, and titanium and its alloys. Nitinol, the shape memory metal, is also beginning to find applications, due to its unusual properties, as discussed below. Small quantities of silver and noble metals, such as platinum and gold, also have their uses in biomedical applications due to their inert behaviour.

2.2   The metallic bond

All atoms consist of a small nucleus of neutrons and protons surrounded by orbiting electrons. Different atomic models can be used to describe different properties of the overall material. One useful model of a metal is a repeating structure of metallic ions surrounded by a sea or cloud of electrons (CD Fig. 2.2). These electrons are not bound to any particular ion but are free to move around within the structure. It is these 'free electrons' that result in metals being good conductors of heat and electricity, opaque to visible light, and that result in the familiar metallic lustre of a polished metallic surface. Alternatively, many of the mechanical properties of metals can be better understood by considering the atoms to behave like hard spheres. These spheres are bonded together by an attractive force. The bonding is non-, or only very weakly, directional and, as a result, metals have a crystalline structure in which the atoms are arranged in a relatively dense, regular, repeating pattern. The atoms of different metals are arranged in different crystal structures. Examples include aluminium in which the atoms are arranged in a face centred cubic (f.c.c.) structure (CD Fig. 2.3 and 2.4) and titanium in which the atoms are arranged in a hexagonal close packed (h.c.p.) structure (CD Fig. 2.5 and 2.6). At room temperature iron has the third, most common, crystal structure in metals that is body centred cubic (b.c.c., CD Fig. 2.7). Many physical and mechanical properties are determined by the strength of the metallic bonds and the arrangement of the atoms in the crystal structure. For example, when a metallic material melts, the atoms have to gain sufficient energy to break free from the crystal structure. If the bonds are numerous and strong, the melting point will be high. Another physical property determined by the crystal structure is the density. The density of a metal is its mass divided by the volume that it occupies. The mass of each atom is fixed for any particular element but the number of atoms in a fixed volume is determined by the crystal structure. The crystal structure also determines how atoms are able to slip over one another when a metal is deformed. If it is difficult for the atoms to slide over one another, the metal is likely to break and is brittle, if the atoms slide easily over one another the metal can deform and it is ductile.

2.3   Microstructure

Metals are rarely used as single crystals but more commonly are polycrystalline and may also be mixtures of two or more different phases. Each phase is physically and chemically distinguishable from the next. It might have a different crystal structure or a different composition. The arrangement of the crystals (or grains) and of the different phases is called the microstructure of the material. The microstructure can often be easily observed by using an optical microscope (CD Fig 2.8). Features as small as 1 μm can be observed in this way, which is equivalent to a magnification of 2000 ×. One particular phase, or a boundary between the phases, is usually highlighted by chemically staining or dissolution (etching). Smaller features can be observed using electron microscopes. Scanning electron microscopes enable magnifications of up to 40 000 × with a large depth of field, ideal for observing fracture surfaces (CD Fig. 2.9). It is possible to see further detail using a transmission electron microscope with magnifications of up to 100 000 × but these images must be taken through thin foils of the sample (CD Fig. 2.10).

The microstructure of any material is important, as it affects many of the mechanical, and some physical properties. For example, at ambient temperature the grain size of a metallic material has a strong effect on the yield stress σ_{yield} at which it will begin to undergo permanent (plastic) deformation. This can be described mathematically by the Hall–Petch equation.

$$\sigma_{yield} = \sigma_0 + Kd^{-1/2} \tag{2.1}$$

where d is the average grain diameter and σ_0 and K are constants for the material. This relationship is shown graphically for a titanium alloy in CD Fig. 2.11.

The grain size can be controlled by the processing conditions. These include changes in the rate of solidification, and by controlled cold deformation and heat treatments. Alterations to the chemical composition, and other processing conditions, can also have dramatic effects on the microstructure and, therefore, the properties of the materials being produced.

2.4   Mechanical properties

When a force is applied to a material, an opposing force exists within the material. This opposing force is measured as a force per unit area, giving the stress in the material. Stress has units of force/area or in SI, pascals (Pa). The stress causes the material to deform. The ratio of the deformation to the original dimension is the strain. Strain does not have units as it is a ratio. If a metallic material is deformed by applying a tensile stress, elongation occurs in the direction of the force (Fig. 2.1). The strain is the ratio of increase in length to the initial length of the sample. Initially, when most materials are stressed in this way, there is a linear relationship between the stress and the strain. The gradient of the graph of stress–strain is called the elastic modulus or Young's modulus and this is a material property. If the stress is removed during this stage of the deformation, the material will return, instantaneously, to its original dimensions. This reversible behaviour is called elastic deformation.

Many materials (e.g. ceramics and glass, see Chapter 3) break immediately after they have undergone a critical amount of elastic deformation; however, metals begin to undergo a second stage of deformation called plastic deformation. At this point, the gradient of the stress–strain graph changes and the stress at this point is called the yield stress. This plastic deformation is permanent and, if the metal were to be unloaded, it would retain the strain that had been imparted to it during the plastic deformation.

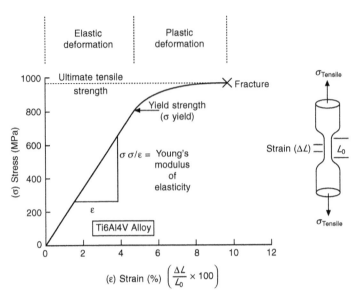

2.1 Stress–strain graph of an orthopaedic titanium alloy tested in tension.

Eventually, as the deformation continues, the metal will fracture. The maximum tensile stress that can be applied to a material before it fractures is called the tensile strength of the material. Usually a metal has deformed so much at this point that it has already failed to perform its function, so, when selecting metals for a given application it is usually the yield strength, and not the tensile strength that is important.

There are other important properties that can be read from this type of graph of stress against strain. These include the elongation or the ductility of the metal. This is a measure of the amount of strain a metal can undergo before fracture. It is commonly expressed in various ways. The plastic strain to failure might be quoted, or the percentage change in length of the failed specimen to its original length (in this case the original length should also be noted). Reasonable levels of ductility are vital for efficient processing of the material. Toughness is a measure of the resistance of a material to crack propagation and one way of measuring toughness is as the area under the stress–strain graph. This area is the energy required to fracture the specimen. If the energy is high, the material is difficult to fracture and it is said to be tough. Conversely, a low energy is found for a brittle material; such a material is relatively easy to fracture and will often fail with little or no warning. Tough materials are also more resistant to sudden impact loads. Typical values of these properties for selection of metallic materials are given in the table in CD Fig. 2.13 and in Chapter 12.

Metals are the material of choice for orthopaedic applications because of their high yield strength and toughness. High and sudden stresses can be encountered in the body in orthopaedic applications and the metal must withstand these stresses without undergoing permanent deformation or fracture. Ideally, we would also select a material with an elastic modulus similar to that of bone. This would ensure the best possible joining and least damage to the parts of the bone where the metal is attached, and the avoidance of stress shielding effects, which are discussed in subsequent chapters.

2.5   Fatigue properties

Often, if a material undergoes some type of cyclic loading, it will fail at a stress much lower than the yield stress under a simple single cycle test. This effect is called fatigue. There are two types of behaviour that occur during fatigue. Most alloys will fail at a decreasing cyclic stress amplitude as the number of stress cycles increases (CD Fig. 2.14). For these materials, the fatigue life is defined as the number of cycles a material will survive at a given stress amplitude. Alternatively, the fatigue stress might be quoted: this is the stress that a material will survive for some large number of cycles (10^7 or 10^8 cycles are typical). Aluminium, magnesium and most non-ferrous (not iron-based) metals show this type of behaviour. For aluminium the fatigue strength is typically one-third of the tensile strength of the alloy. Many steels and titanium alloys instead show a fatigue limit. Below a certain stress amplitude (the fatigue limit) the alloy will show a virtually indefinite fatigue life without failure. This is typically at about half of the tensile strength of the material.

Unfortunately fatigue is a common cause of failure in metallic components. It can often be easily identified by characteristic striations on the fracture surface. These are microscopic lines that form during each loading cycle. Although the presence of such markings confirms that the component has undergone fatigue, their absence does not preclude fatigue failure, as they often become obliterated during subsequent loading cycles. A corrosive environment (such as a saline solution) can often have serious detrimental effects on the fatigue properties of a metallic material and this is a serious selection problem for biomedical applications due to the corrosive nature of body fluids (see below).

2.6   Hardness and wear

The hardness of a metal is its resistance to localised plastic deformation by indentation or scratching and is one of the easiest mechanical properties to measure. A small indentor is pressed into the surface of the metal and the size of the indent formed is measured to calculate a hardness value. Different hardness scales are used with the indentor made from different hard materials (such as hardened steel or diamonds), cut in different shapes (e.g. spheres or pyramids). As well as measuring the resistance to indentation or scratching of a metal surface, the hardness values can be used in quality control and to estimate other mechanical properties such as the tensile strength of the alloy. The hardness of a metal also influences its wear behaviour when there is relative motion between two surfaces. The wear will also depend on the presence of any lubrication, the roughness of the sliding surfaces, the chemical environment and the loading conditions. Wear is important not only due to the damage caused to the component but also due to the generation of undesirable debris.

2.7   Shape memory effect and superelasticity

The amazing ability of some materials to 'remember' a previous shape was first found in the 1950s in a gold–cadmium alloy. The alloy can be deformed at a low temperature and will return to its original shape when heated to a critical temperature. The most important practical shape memory alloys are nickel–titanium (Ni–Ti) alloys, often called by the name 'nitinol'. These alloys show both the shape memory effect and a related property called superelasticity. Both effects are useful in biomedical applications.

The shape memory effect (to be abbreviated to SME) relies on a solid-state change in the crystal structure of the alloy. The parent crystal phase exists at high temperature and a 'martensitic' phase exists at low temperatures. As the crystal structure changes between the two temperatures, the atoms move in a very controlled way. When the alloy is heated above the transformation temperature for the parent phase, the atoms return to their original positions. This behaviour is quite unusual and only occurs in some materials. Consider the 2-D schematic of the crystal behaviour in CD Fig. 2.15. In the high temperature phase the atoms are arranged in a square structure. The low temperature phase (martensite) has a lower symmetry with the atoms arranged in a diamond pattern. Due to this, different orientations of the same martensitic phase can form when the material is cooled; these are called 'correspondence variants'.

SME occurs when the low temperature martensitic phase is deformed. One of the correspondence variants will grow at the expense of the others. Upon heating, all of the variants revert to the parent phase in the original square structure by the reverse phase transformation and the deformation is reversed.

Ni–Ti has many, relatively new, medical applications. The shape memory effect is used in bone plates to provide a compressive force on the fracture, resulting in faster healing. Marrow needles (used for fixation of a broken thigh bone) are difficult to insert but this can be made easier with a shape memory alloy because the required size and shape will recover after heating the needle in the marrow. An Ni–Ti shape memory alloy wire has also been used in an artificial kidney pump. The Ni–Ti is used as an actuator, an electric current causes it to heat up and apply a force, compressing the pump. The current is removed and the pump expands as rapid cooling of the thin wire causes it to transform to martensite, which is easily deformed by a bias spring. The Ni–Ti wire can fulfil the demands of miniaturisation and survive a huge number of cycles.

A related effect, superelasticity, occurs when the transformation to martensite is induced by an applied stress. The correspondent variant that causes the maximum strain for the applied stress forms. When the stress is removed, the Ni–Ti transforms back into its high temperature phase. The material can undergo large amounts of deformation but will spring back to its original shape. This effect was only discovered in Ni–Ti in 1981. The mechanical properties are ideal for the mechanical manipulation of misaligned teeth in orthodontics and result in faster movement of the teeth and reduced need for readjustment when compared with more conventional stainless steel wire.

Other applications in orthodontics include utilising the shape memory effect in blade implants to attach teeth and clasps in attaching partial dentures. The force applied by the superelastic wire, the deformation achieved in the shape memory effect and the transformation temperatures can be controlled by the processing routes employed and by the microstructures obtained in forming the metal.

2.8   Corrosion

The corrosion of metals in the body may cause the component to break or fail to perform its function and may also create harmful corrosion products. All metals, except the noble metals (e.g. gold (Au) and platinum (Pt)) are prone to corrosion. The corrosion of metals occurs by electrochemical attack, where electrons are transferred from one chemical species to another. The corroding metal loses electrons to become a metal ion and is said to be oxidised (even if oxygen is not involved in the process). The site at which the oxidation reaction occurs is called the anode. These electrons are consumed in a reduction reaction, which occurs at the cathode.

Body fluid has corrosion effects on metals very similar to warm aerated seawater. Metal ions can dissolve and diffuse through the fluid, and electrons are consumed by reaction with either dissolved oxygen in the water to form hydroxyl (OH^-) ions or by combination with positively charged metal ions.

Although the corrosion process is always electrochemical, several different mechanisms may be occurring and these types of corrosion can be easily recognised. Some of these mechanisms include galvanic corrosion, crevice corrosion, pitting and stress corrosion. The different types of corrosion can be prevented in different ways, although preventing contact between the metal and the corrosive environment using a carefully chosen coating will usually work. To determine other methods to prevent corrosion, it is necessary to have some understanding of how the different types of corrosion occur.

Galvanic corrosion is a particular type of corrosion that occurs when two dissimilar metals are electrically connected in a corrosive environment, e.g. two metals in contact in a liquid electrolyte such as body fluid. The corrosion rate of one of the metals is increased and that of the other metal is decreased compared with when they are not in electrical contact with each other. The metal that corrodes faster becomes the anode and the other the cathode and an electrical potential is set up between them. (This is the same way in which dry cell batteries work.) Galvanic corrosion can be prevented by insulating the metals electrically from one another, or by carefully selecting combinations of metals that do not strongly affect the corrosion rates.

Crevice corrosion is a highly localised form of corrosion that occurs within crevices and shielded areas, commonly at joints or under gaskets. The remaining surfaces show no, or very little, damage. The crevice is usually hundredths of a millimetre or less in size. This is wide enough to permit liquid entry but narrow enough to maintain a stagnant zone. Stainless steels are particularly susceptible to this type of corrosion and can be cut by the stagnant zone formed under an elastic band in seawater. Initially, corrosion occurs uniformly over the entire surface of the component. After a short time, the oxygen within the crevice becomes depleted because of the stagnant liquid. No reduction of oxygen to hydroxyl ions occurs in this area, although the dissolution of metal continues. An excess of positive charge is produced in the solution by the metal ions. This is balanced by the migration of chloride ions into the crevice. The concentration of metal chloride increases within the crevice. Metal salts hydrolyse in water forming an insoluble metal hydroxide, which is deposited, and a free acid, following the reaction below:

$$M^+Cl^- + H_2O \rightarrow MOH\downarrow + H^+Cl^-$$

Both chloride and hydrogen ions accelerate the dissolution rates of most metals and alloys and the process becomes autocatalytic. The mechanism is shown in CD Fig. 2.16. Careful design to avoid crevices can prevent this type of corrosion.

Pitting occurs by a very similar method to crevice corrosion and can be recognised when narrow pits grow, usually down from a horizontal surface. The pits may initiate at a scratch or defect in the material or due to random variations in the concentration of the corrosion fluid. Again, stainless steels are susceptible to pitting, but alloying can reduce it. For example 2% Mo can be added to 18–8 stainless steel to prevent pitting. Certain materials are known not to pit in specific environments, for example titanium is generally resistant to this type of corrosion.

Stress corrosion cracking results from the combined action of a tensile stress and a corrosive environment. Both of these conditions are necessary; some materials may hardly corrode at all in the corrosive environment unless a stress is also present. Small cracks form and propagate perpendicular to the stress, eventually leading to failure. The stress need not be an externally applied stress and may be due to the processing conditions or due to the presence of second phases in a metal alloy. Most alloys are susceptible to this type of corrosion in a specific environment, for example stainless steels stress corrode in acid chloride solutions and titanium alloys stress corrode in seawater. Fortunately, the number of different environments in which a specific alloy will crack is small. The best way of reducing or preventing this type of corrosion is to reduce the stress that the metal is subjected to. Alternatively, the metal can be separated from the corrosive environment by a coating, or a different alloy can be used.

2.9   Effects of processing on structure, properties and reliability

The interrelations between composition, processing, microstructure and properties cannot be over-emphasised. The microstructure and composition of a material determine many of its physical and chemical properties and all of its mechanical properties. The microstructure is determined by the processing conditions and the composition. However, the evolution of the microstructure during processing may limit the processing operations that can be performed. The material may become too hard or brittle for the processing to continue. Alternatively, structures that may result in poor properties of the finished component may form in the work piece.

Processing techniques include the solidification conditions, and subsequent deformation and heat treatments. It has already been mentioned how grain size can influence the yield stress in metallic alloys, as described by the Hall–Petch equation. Grains can also become aligned in certain crystallographic directions during processing, which will also affect the material properties. Deformation introduces defects into the metallic crystal structures, which generally improve mechanical properties such as yield strength but are deleterious to properties such as the ductility. Heat treatments to the metals and alloys can increase the grain size, remove defects or result in second-phase particles dissolving or precipitating out in to the material. Controlling the cooling rate of a material can result in its transformation into completely new phases with very different properties. Defects such as scratches or pores may appear in the material that will have a deleterious effect on its properties.

2.10   Clinical requirements

For a metallic material to be used as a biomaterial, it must have adequate physical and mechanical properties to perform its function. It must not corrode significantly in the body and it must last for the required lifetime whether determined by the number of cycles that the component must withstand or the time that a component must survive inside the human body. Metals are often chosen for a combination of their strength and toughness (stainless steels, titanium alloys, cobalt alloys, all used in orthopaedic applications), their electrical conductivity and their inert behaviour (gold, platinum) or novel behaviour such as shape memory and superelasticity (Ni–Ti, used in orthodontics, orthopaedics and miniature devices). See Chapter 12 for a summary of composition and properties of alloys used in medical applications.

2.11   Summary

This chapter began by introducing the reader to metallic materials and described some of their distinguishing properties due to the nature of the metallic bond. The concept of crystal structure within a metal and the existence of a microstructure within a metallic material was then introduced. The definition and measurement of a wide range of mechanical properties including elastic modulus, yield stress, strength, ductility and toughness, fatigue, hardness and wear was described. The shape memory effect and superelasticity were described and some useful biomedical applications of these effects were discussed. The electrochemical nature of metallic corrosion was described and four specific corrosion mechanisms, pertinent to biomedical applications, were discussed, namely galvanic corrosion, crevice corrosion, pitting and stress corrosion. The influence of processing conditions on the microstructure and hence the properties of the metallic materials was emphasised and briefly exemplified. Finally, general clinical requirements and examples of metals used in biomedical applications were mentioned. Space has only allowed a very general introduction to metals to be given and readers are encouraged to expand their knowledge initially with the aid of the further reading list below.

2.12   Reading list

General introduction to important metallic materials

Alexander W. and Street A., *Metals in the Service of Man*, 11th edition, New York, Penguin, 1998.

Basic principles of materials science including aspects related to metals and alloys

John V., *Introduction to Engineering Materials*, 3rd edition, New York, Palgrave, 1992.
Ashby M.F. and Jones D.R.H., *Engineering Materials 2*, 2nd edition, Oxford, Butterworth Heineman, 2001.
Callister W.D., *Materials Science and Engineering: An Introduction*, 6th edition, New York, Wiley, 2003.

Mechanical properties and processing of metals

Dieter G.E., *Mechanical Metallurgy*, SI metric edition, London, McGraw Hill, 1988.

Shape memory effect and superelasticity with a selection of medical (and other) applications

Otsuka K. and Wayman C.M., *Shape Memory Materials*, Cambridge, Cambridge University Press, 1999.

Corrosion behaviour

Fontana M.G., *Corrosion Engineering* 3rd edition, London, McGrawHill, 1986.

3
Ceramics

ALDO R. BOCCACCINI
Imperial College London, UK

3.1 Introduction

Ceramics are solid materials composed of inorganic, non-metallic substances. This definition includes 'traditional' ceramics such as porcelain, pottery and cements but also 'advanced' or 'engineering' ceramics such as ferroelectrics, non-metallic magnetic materials, bioceramics and structural oxide and non-oxide materials. Ceramics exist as both crystalline and non-crystalline (amorphous) compounds. Glasses and partially crystallised glasses, known as glass-ceramics, are therefore subclasses of ceramics.

A common characteristic of all ceramic materials is that they are subjected to high temperatures during manufacture or use (high temperature usually means above 500 °C). Typically, but not exclusively, a ceramic is a metallic oxide, boride, carbide or nitride, or a mixture or compound of such materials, i.e. it includes anions that play important roles in their atomic structure and properties. Major characteristics of ceramics are their high hardness, insulating properties of heat and electricity, heat and corrosion resistance and their brittleness and fracture behaviour without deformation.

Ceramics used for implants and in the repair and reconstruction of diseased or damaged body parts are called 'bioceramics'. Bioceramics are used mainly to repair hard tissues such as bones, joints or teeth. Many ceramic compositions have been tested for use in the body; however, only few have achieved human clinical application. Examples of bioceramics are alumina, zirconia, titania, tricalcium phosphate, hydroxyapatite, calcium aluminates, bioactive glasses and glass-ceramics. Depending on the type of response in the body, bioceramics can be broadly classified as bioinert, bioactive and resorbable, as discussed further below. Three factors influence the choice of ceramics and glasses as biomaterials: (i) physical and mechanical properties, (ii) degradation of the material in the body, and (iii) biocompatibility. The microstructure of the ceramic material, its properties and the processing routes used are all very strongly interrelated. This chapter introduces the reader to general principles that explain some of the mechanisms behind these factors. These can be applied to a wide variety of ceramics and glasses but there will be a specific emphasis on ceramics with properties and processes relevant to biomedical applications.

3.2   Atomic bonds and atomic arrangements in ceramics

Polycrystalline ceramics are solids in which the atoms or ions are arranged in regular arrays. In glasses (amorphous materials), the regularity (order) is only short-range. The type of atomic bonding and the atomic arrangement affect the properties of the materials. Atomic bonding in ceramics is mainly ionic or covalent, and usually a hybrid of these. The tendency towards ionic bonding increases with increasing difference in the electronegativity of the atoms. In ionic bonding there is a transfer of electrons between the atoms making up the compound. Positively charged ions balance out the negatively charged ions to give an electrically neutral compound. One example is $NaCl$, where Na^+ ions balance the Cl^- ions. In covalent bonding, electrons are shared between atoms. The crystal structure determines how atoms are able to slip over one another when a material is deformed. If it is difficult for the atoms to slide over one another, the material is likely to break and is brittle, if the atoms slide easily over one another, the material can deform and it is ductile. The characteristic brittleness and high strength of ceramics can be traced, therefore, to the ionic and covalent bonding types: they cause a very high inherent resistance to dislocation motion in the lattice and thus to plastic deformation. Therefore, unlike in metals, a stress concentration at a crack tip in a ceramic cannot be relieved by plastic deformation and the material fails in a brittle fashion.

The spatial arrangement of individual atoms in a ceramic (the crystalline structure) depends on the type of bonding, the relative sizes of the atoms and the need to balance the electrostatic charges. CD Figs. 3.5–3.7 show schematics of the unit cells of the types of crystal structures found in ceramics. Most ceramic crystal structures are variants of the face centred cubic (f.c.c., CD Fig. 3.5) or hexagonal close packed (h.c.p., Fig. 3.1 and CD Fig. 3.7) structures. Typically, the metallic cations occupy interstitial positions in a lattice made of non-metallic ions.

The most common crystal structures in ceramics are:

1. simple cubic, e.g. $CsCl$, $CsBr$, CsI;
2. close packed cubic: this is a variant of the face-centred cubic (f.c.c.) structure, e.g. CaO, MgO, FeO, BaO, etc;
3. closed packed hexagonal, e.g. Al_2O_3, Fe_2O_3, Cr_2O_3, etc.

Alumina (Al_2O_3) is the most important crystalline bioceramic as it is used in joint replacements (Chapter 13). Its h.c.p. crystal structure is shown in Fig. 3.1 and in CD Fig 3.7.

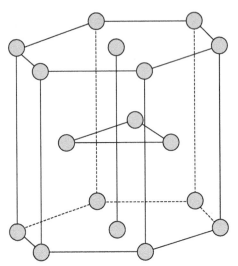

3.1 Unit cell of a hexagonal close packed (h.c.p.) crystal lattice structure (CD Fig. 3.5).

Glasses, as a subclass of ceramics, constitute the most important group of non-crystalline solids. Glasses are amorphous materials of very high viscosity, which behave like solids. Glasses maintain a disordered atomic structure characteristic of a liquid, i.e. they do not undergo transformation to a crystalline state. Silicate glasses are made up of a network of tetrahedra of four large oxygen ions with a silicon ion at the centre (Fig. 3.2, CD Figs. 3.15, 3.16, 19.12, 19.13). Each oxygen is, however, shared by two tetrahedra, giving the bulk composition of SiO_2. The basic silicate network can incorporate virtually

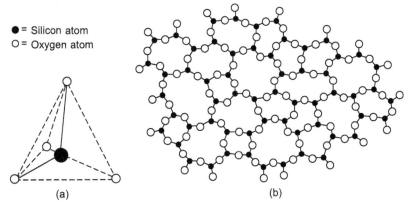

● = Silicon atom
O = Oxygen atom

(a)   (b)

3.2 Glassy structure. (a) A silica tetrahedron and (CD Fig. 3.15, CD Fig. 19.13) (b) a randomised network of silica tetrahedra (CD Fig. 3.16, CD Fig. 19.12).

all atoms of the periodic table of elements, and thus silicate glasses of numerous different compositions and properties can be obtained. Silicate glasses are therefore the most common and technologically the most important group of glasses. One of the major advantages of glasses is the ease of fabrication, which allows processes such as melt infiltration and compression moulding to be used.

Glass-ceramics represent a special class of ceramics. They are polycrystalline materials having fine ceramic crystallites (usually of size < 1 μm) in a glassy matrix. They are produced by the controlled crystallisation of glasses using suitable heat treatments. Glass-ceramics of technical relevance are, for example, in the systems $Li_2O–Al_2O_3–SiO_2$ (LAS), $MgO–Al_2O_3–SiO_2$ (MAS), e.g. cordierite ($2MgO.2Al_2O_3.5SiO_2$) and $SiO_2–Al_2O_3–MgO–K_2O–F$, e.g. fluoromicas.

Figure 3.3(a) (CD Fig. 3.8(b)) summarises the time–temperature profiles used in processing melt-derived glasses and ceramics and glass-ceramics that have been produced by melt processing, sintering or hot pressing. Figure 3.3(b) (CD Fig. 3.9) shows the time-dependent processing steps for sol–gel-derived glass production and compared to melt processing. The thermal processing of sol–gel-derived glasses and ceramics involves much lower temperatures, but melt processing is less time consuming, due to the carefully controlled thermal processing required to create crack-free sol–gel derived glasses.

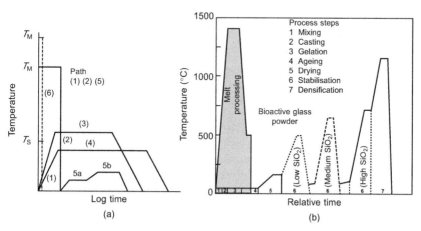

3.3 (a) Time–temperature profiles of processing steps for (1) glass, (2) cast polycrystalline (large grained) ceramic, (3 and 4) solid-state sintered ceramic, (5) polycrystalline glass-ceramic, (6) polycrystalline coating from liquid, T_m is the melting temperature, T_s is the solidus line of the phase diagram (CD Fig. 3.8(b)). (b) Processing steps for sol–gel-derived glass production compared to melt processing (CD Fig. 3.9).

3.3   Microstructure of ceramics

Ceramics are rarely used as single crystals. They are more commonly polycrystalline and may be mixtures of two or more different crystalline phases. Sometimes, a glassy phase is also present. Each phase is physically and chemically distinguishable from the next. Different phases vary in crystal structure and/or composition. The arrangement of the crystals (grains) and of the different phases constitutes the microstructure of the material. The microstructure depends on the initial fabrication techniques, raw materials used, phase changes, chemical reactions and grain growth occurring during the high temperature processing (see Section 3.5). In general, ceramic microstructures consist of individual crystals (grains) separated by boundary layers, often containing a glassy phase, the grain boundaries, and gas filled pores. A wide range of grain sizes is observed, generally 1–1000 µm. Porosity may be fine or coarse, open or closed. Ceramic microstructures can be observed after polishing the materials by using optical microscopes. One particular phase, or the grain boundaries, may be highlighted by chemical or thermal etching. Features smaller than ~ 1 µm on polished sections and fracture surfaces can be observed using electron microscopes.

The microstructure of any material is important as it affects the material's mechanical and physical properties. In particular, the grain size and shape, grain size distribution, nature of grain boundaries and pore structure are very important parameters. In contrast to polycrystalline ceramics, there are no microstructural features in glasses, which appear homogeneous under optical or electron microscopes, whereas glass-ceramics microstructures are characterised by a dispersion of crystals, usually of micrometre size, dispersed in a continuous glassy matrix.

3.4   Mechanical properties of ceramics

Elastic modulus, fracture strength, fracture toughness

When a force is applied to a material, an opposing force exists within the material. This opposing force is measured as a force per unit area giving the stress in the material. The stress causes the material to deform. The ratio of the deformation to the original dimension is the strain (CD Fig. 3.19). Initially, there is a linear relationship between the stress and the strain (Fig. 3.4). The gradient of the stress–strain dependence is the elastic modulus or Young's modulus and this is a material property. If the stress is removed during this (initial) stage of the deformation, the material will return, instantaneously, to its original dimensions. This reversible behaviour is called elastic deformation. The elastic properties of ceramics determine their mechanical behaviour and are closely related to the crystal structure and atomic bonding. Elastic constants are also strongly dependent on the microstructure. In particular, porosity has

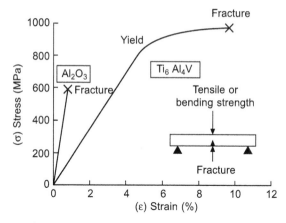

3.4 Stress–strain graph for an alumina ceramic compared with a metal alloy.

a significant effect on elastic constants, which decrease with increasing pore content. Besides pore content, pore shape and pore orientation also affect the elastic behaviour of ceramics and glasses.

Ceramics and glasses break immediately after they have undergone a critical amount of elastic deformation; metals, however, begin to undergo a second stage of plastic deformation before fracture (see Chapter 2). The maximum tensile stress that can be applied to a material before it fractures is called the tensile strength of the material. The common strengths of ceramics are one or two orders of magnitude less than the theoretical values calculated from interatomic forces. This is because the fracture of ceramics and glasses normally occurs by the propagation of cracks or flaws existing in the materials, as discovered by Griffith in 1920. These cracks have a stress concentrating effect, which weakens the material. The ease with which a sheet glass can be 'cut' after light scribing with a sharp diamond tip is a demonstration of this effect. The concentrated stress at the tip of a narrow crack is inversely proportional to the radius of curvature at the crack tip; thus sharp cracks are more detrimental to tensile strength than rounded flaws.

Typical flaws in ceramics are:

1. processing-induced flaws, such as inclusions, pores, isolated large grains, glassy grain boundaries and machining-induced surface cracks;
2. design-induced flaws, such as sharp corners, holes, etc;
3. service-induced flaws, mainly due to environmental degradation, thermal stresses, impact loads and wear.

Porosity is one of the most common defects in ceramics, which also negatively affects strength. Careful processing can eliminate many flaws or reduce the flaw size. The size of the crack strongly affects the strength of the material.

The fracture strength (σ) of a ceramic or glass material containing an internal crack of length $2c$ is given by the Griffith's relationship:

$$\sigma = (2E\ \gamma/c)^{1/2} \tag{3.1}$$

where E is the Young's modulus and γ is the surface energy of the crack surface.

There are other important properties that can be read from the graph of stress against strain. Toughness is a measure of the resistance of a material to crack propagation and one way of measuring toughness is as the area under the stress–strain graph. This area is proportional to the energy required to fracture the specimen. A low energy is found for brittle materials, such as ceramics and glasses, which often fail with little or no warning. The relationship between fracture toughness (K_{Ic}) and tensile strength in a ceramic containing a crack of length $2c$ is given by the following relationship:

$$K_{Ic} = \sigma Y c^{1/2} \tag{3.2}$$

where Y is a dimensionless constant which depends on the geometry of the loading and the crack configuration.

The incorporation of particles or fibres into ceramic and glass matrices, forming composite materials, is a common method used to increase the toughness of these brittle materials (Chapter 5). Ceramic composites are also more resistant to sudden impact loads. The low fracture toughness of ceramics and glasses is the greatest impediment to their broader use in load-bearing orthopaedic applications in the human body. However, bioceramics find broad applications as coatings on metallic implants in load-bearing orthopaedic applications and in a variety of other areas in the muscoskeletal and dental fields not involving high loads.

Subcritical crack growth

Time under stress is a very important parameter when considering the strength of ceramics. The strength of ceramics under constant load changes with time. The time dependence of strength is due to subcritical crack growth occurring under stress, which is often assisted by environmental factors, such as water from the environment. Below a certain value of stress intensity factor, K_{I0}, no crack growth occurs. At higher values of K_I, subcritical crack growth occurs at a constant rate, which depends mainly on the diffusion of corrosive species from the environment (e.g. body fluids in case of a bioceramic) to the crack tip. Safe operation of ceramic components is under conditions leading to stress intensity factors below K_{I0}.

Hardness and wear

The hardness of a material is a measure of its resistance to localised deformation by indentation or scratching and is one of the easiest mechanical properties to measure. A small indentor is pressed into the surface of the material and the size of the indent formed is measured to calculate a hardness value. The hardness and fracture toughness of a ceramic influence its wear behaviour when there is relative motion between two surfaces. The wear will also depend on the presence of any lubrication, the roughness of the sliding surfaces, the chemical environment and the loading conditions. Wear is important not only due to the damage caused to the component but also due to the generation of undesirable debris, which, in the case of bioceramics, can lead to failure of the component (implant) and/or inflammatory responses in the surrounding tissue.

3.5   Processing of ceramics

There are several methods for producing glasses and ceramics. The following steps are involved in the conventional processing of ceramic components:

1. The starting material is usually in the form of a powder, slurry or a colloidal suspension.
2. A preliminary component is formed, which is called a green body. The green body has the shape of the final component but is generally of lower strength. In the case of powders, this is done by pressing to produce a powder compact. In the case of colloidal suspensions and slurries, several methods can be used, e.g. tape and slip casting, injection moulding, pressure filtration and electrophoretic deposition.
3. The green body is heat treated (sintered) to increase density and strength of the component. Frequently, a combination of high temperature and pressure (hot-pressing and hot isostatic pressing) is used.
4. The sintered dense compact is machined to the final required shape.

The conventional processing of glass involves the melting of raw materials, followed by casting; however, several products for biomedical applications, made, for example, of bioactive glasses, are prepared by the 'ceramic powder methods' described above. Sintering is the most important process used for consolidating ceramic and glass powders. This technique involves atomic diffusion processes in the case of polycrystalline ceramics and viscous flow in the case of amorphous glasses. Other techniques used for producing ceramic and glass components involve chemical routes, which usually lead to a high degree of homogeneity on a molecular scale and consequently high purity products can be obtained. The sol–gel process belongs to this category. The sol–gel process has extensive application for production of porous and foam-like bioactive glasses, ceramics and ceramic coatings. This process involves the hydrolysis of silicon alkoxide in solution to form a colloidal solution (sol) and the subsequent chemical polymerisation (condensation reaction) of the silica units to form gel. The gel is heat treated to dry it and form a glass or ceramic structure at temperatures much lower than those used for powder sintering.

3.6   Impact of fabrication on microstructure and properties

The microstructure and composition of a material determines many of its physical, and all of its mechanical, properties. The microstructure is determined by the processing conditions and the composition. The evolution of the microstructure during processing (for example, densification during sintering) may limit the processing operations that can be performed. Alternatively, microstructures that may result in poor properties of the finished component may form in the work piece. The complex interaction between fabrication, microstructure, mechanical properties and external factors (e.g. temperature, environment conditions) is given in Fig. 3.5.

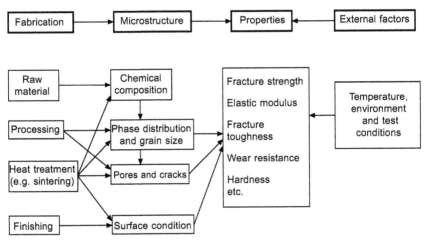

3.5 The interaction between fabrication conditions, microstructure, mechanical properties and external factors in ceramic materials (modified from Davidge, 1979).

3.7   Clinical requirements

Bioceramics are used mainly for repair and reconstruction of diseased or damaged parts of the musculoskeletal system. The choice of a particular bioceramic for a given application will depend on the type of bioceramic/ tissue attachment required. Dense, non-porous, almost inert ceramics such as Al_2O_3 and ZrO_2 attach by bone growth into surface irregularities by cementing the device into the tissue, or by press-fitting into a defect (mechanical fixation). If these implants have pores with diameters in excess of 100 μm, bone ingrowth can occur, which anchors the bone to the material (biological fixation). In surface reactive ceramics, such as hydroxyapatite and certain compositions of silicate glasses and glass-ceramics are used, the materials attach directly by chemical bonding with the bone (bioactive fixation). Bioactive ceramics are also used as coatings on metallic implants. Resorbable ceramics and glasses in bulk or powder form are designed to resorb in the body at the similar rate of formation of new bone. Some bioactive glass compositions, particularly in the system SiO_2–CaO–Na_2O–P_2O_5 also bond to soft tissues. Bioceramics, therefore, have high potential to be used as porous scaffolds for tissue engineering applications (Chapter 19).

For a ceramic material to be used as a biomaterial, it must have adequate physical, biological and mechanical properties to perform its function. The major challenge facing the use of ceramics in the body as permanent implants is to replace old, deteriorating bone with a material that can function for the remaining years of the patient's life. Survivability of a bioceramic requires formation of a stable interface with living tissue. If interfacial movement can occur, the implant loosens rapidly. Loosening invariably leads to clinical failure, in the form of fracture of the implant or the bone adjacent to the implant. In case of using (almost) inert bioceramics, bone at the interface is very often structurally weak because of disease, localised death of bone, or stress shielding that occurs because the higher elastic modulus of the implant prevents the bone from being loaded properly. The attachment and bonding of (almost) inert implants (both metallic and ceramic) to bone can be enhanced by using designed porous structures or by using bioactive ceramic materials, such as hydroxyapatite or bioactive glasses, as coatings. The potential advantage offered by porous ceramics is the mechanical stability of the interface that develops when bone grows into the pores. Lower mechanical strength and brittleness are the disadvantages associated with porous ceramic implants that have restricted their uses primarily to non-load-bearing applications.

Resorbable bioceramics are designed to degrade gradually over time and be replaced by the natural host tissue. In fact, this could be seen as the optimal solution to biomaterials problems. However, developments of resorbable ceramics present complications, mainly due to problems related to matching the rate of resorption with the replacement by the natural host tissue and maintenance of strength and stability of interfaces during the degradation period. The rate of tissue growth varies from patient to patient and with tissue type.

Another approach to the solution of the problems of interfacial attachment is the use of bioactive materials. These are materials that elicit a specific biological response at the interface of the material, which results in the formation of a bond between the tissues and the material. As mentioned above, hydroxyapatite and selected compositions of silicate glasses (e.g. Bioglass®) and apatite–wollastonite glass-ceramics are bioactive. In fact, bioceramics that are both resorbable and bioactive constitute the 'third-generation' biomaterials, which find applications as scaffolds in osseous and soft tissue engineering. In these applications, the bioactive ceramic can be formed into a porous (foam-like) structure with tailored pore size and orientation or it can be used as a filler or coating in resorbable polymers, forming optimal scaffolds exhibiting both resorbability and bioactivity (Chapter 19).

The mechanical properties of bioceramics, in particular their low fracture toughness are disadvantages for their direct use as bone replacement in load-bearing applications. Usually, bioceramics are combined with polymers and metals effectively forming composite materials with enhanced mechanical properties and elastic constants matched to those of bone.

3.8   Summary

This chapter has introduced the reader to ceramics (and glasses), describing some of their distinguishing properties due to the nature of the atomic bonds and, in the case of polycrystalline ceramics, their crystalline structure. The concept of a microstructure within a ceramic material was introduced, paying attention to grain size, porosity and flaws existing in all ceramics used in practical applications. The definition and measurement of several mechanical properties including elastic modulus, fracture strength, fracture toughness, subcritical crack growth, hardness and wear were described, focussing on the brittle and flaw-sensitive nature of the ceramic fracture mode. Processing methods for ceramics were discussed and the influence of processing conditions on the microstructure and hence the properties of ceramic materials was emphasised. Finally, general clinical requirements and examples of ceramics used in biomedical applications were mentioned.

3.9   Reading list

Chawla K.K., *Ceramic Matrix Composites*, London, Kluwer Academic Publishers, 2003.

Davidge R.W., *Mechanical Behaviour of Ceramics*, Cambridge, Cambridge University Press, 1979.

Green D.J., *An Introduction to the Mechanical Properties of Ceramics*, Cambridge, Cambridge University Press, 1998.

Hench L.L. and Wilson J., *An Introduction to Bioceramics*, Singapore, World Scientific, 1993.

Kingery W.D., *Introduction to Ceramics*, New York, J. Wiley & Sons, 1976.

Rice R.W., *Mechanical Properties of Ceramics and Composites: Grain Size and Particle Effects*, New York, Marcel Dekker, 2000.

4

Polymers

R O B E R T G. H I L L
Imperial College London, UK

4.1   Introduction

Polymers are used widely for medical devices and implants. Early medical devices were based on high purity grades of commonly used industrial polymers. In recent years new polymers have been specially synthesised for medical uses. Table 4.1 gives some common examples of synthetic polymers used in healthcare and their applications.

Table 4.1 Some common polymers and their applications in healthcare

Poly(methylmethacrylate)	Hard contact lenses, intraocular lenses, bone cements, denture base
UHMWPE (Ultra-high molecular weight polyethylene)	Bearing surfaces in artificial joints
PET(Polyethylene terephthalate)	Artificial arteries
Polyurethanes	Catheters
Polyhema (Polyhydroxyethylmethacrylate)	Soft contact lenses, wound dressings, drug release matrices
Poly(propylene)	Sutures, heart valves, finger joints
Silicones	Breast implants, facial devices
Poly(glycolide)	Biodegradable sutures

The term polymer is composed of two terms: 'poly' meaning many and 'mer' meaning unit, hence a polymer is molecule made up of many units. Polymers can be classified into two distinct types:

- thermoset;
- thermoplastic.

Thermoplastic polymers can be moulded to shape by the application of heat and pressure in the molten shape and can be reshaped once formed. This type of polymer is not cross-linked but generally consists of linear polymer chains. Thermoset polymers are generally polymerised in their final state and cannot be reshaped by applying heat. The chains in this type of polymer are generally covalently cross-linked.

Polymers may also be classified as being:

- addition polymers;
- condensation polymers.

Addition polymers are produced by free radical addition reactions from unsaturated monomers containing carbon–carbon double bonds. Examples of addition polymers are polyethylene and poly(methylmethacrylate). Condensation polymers are formed by reacting two monomers together in a reaction in which a small molecule is eliminated, which is often water. Examples of condensation polymers are polyamides and polyesters. Some condensation polymers may be hydrolysed in the body and undergo degradation.

Polymerization reactions result in a distribution of chain lengths and therefore a distribution of molar mass (MW). The exception is many biological polymers, for which MW is strongly controlled. It is assumed that the distribution of MW can be broken down into the number of chains in the mixture that each have a defined length.

The number average molecular mass is determined by techniques that count molecules. It is given by:

$$M_n = \Sigma \, (N_i * M_i / N) \qquad [4.1]$$

where: N_i = the number of molecules of mass i
N = the total number of molecules
M_i = the mass of molecules of length i.

The weight average molar mass is given by:

$$M_w = \Sigma \, (w_i \, M_i) \qquad [4.2]$$

where: w_i = mass fraction

$$w_i = N_i \, M_i / \Sigma \, (N_i \, M_i) \qquad [4.3]$$

The degree of polymerisation indicates the average number of mers per chain and can be defined in terms of either number average or weight average:

$$N = M_w / M_{mer} \quad \text{or} \quad M_n / M_{mer} \qquad [4.4]$$

where M_{mer} is the molecular weight of the mer unit.

The polydispersity is given by:

$$PD = M_w / M_n \qquad [4.5]$$

Typical thermoplastics consist of 10 000 mers, giving rise to long chains of high molar mass. Fully extended chains have a high length to diameter aspect ratio. A typical end to end distance $\approx 2 \times 10^{-6}$ m for a polymer with 10 000 mer units. Freedom of rotation of chemical bonds gives rise to a random coil conformation.

Polymers may be regarded as being like a mass of tangled string or a bowlful of spaghetti.

A polymer may consist of a single mer unit, in which case it is termed a homopolymer. A copolymer consists of two monomer types. The two monomers (A and B) may be in various sequences:

- random – ABABBAABABAABABAABA;
- alternating – ABABABABAB;
- block – AAAAAAABBBBBBBAAAAAA.

Polymers may be amorphous, having no crystallinity, and exhibit glass-like properties below their glass transition temperature and elastomeric properties above the glass transition. Alternatively, they may be semi-crystalline in which case they contain amorphous and crystalline regions (Fig. 4.1, CD Fig. 4.1). A single polymer chain will take part in both the amorphous and the crystalline parts. The crystalline regions act as cross-linked sites, restricting flow, creep and plastic deformation. The degree of crystallinity influences mechanical properties (moduli, strength, etc.) as well gas permeability, water uptake, etc. A typical stress–strain curve for a polymer, such as high density polyethylene, is shown in Fig. 4.2 (CD Fig. 4.2).

Crystalline polymers differ significantly from other materials in that they cannot be obtained 100% crystalline. They do not exhibit a sharp melting point (T_m), but exhibit a range of temperatures over which they melt. They

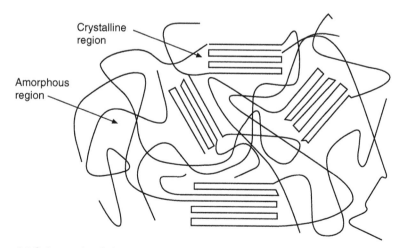

4.1 Schematic of the structure of a semi-crystalline polymer (CD Fig. 4.1, 19.2).

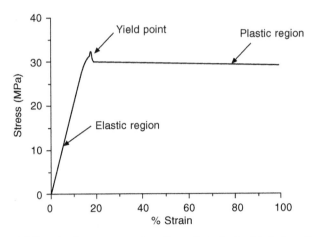

4.2 Graph of stress as a function of strain for high density polyethylene (CD Fig. 4.2).

do not exhibit well-defined latent heats of fusion. They generally exist as small sub-micron crystallites giving rise to line broadening by X-ray diffraction.

Crystallinity influences many properties. Diffusion of small molecules can often only take place through the amorphous regions. As a consequence diffusion reduces with an increase in crystallinity. In order to dissolve readily, a semi-crystalline polymer has to be at a temperature above T_m. The Young's modulus and yield stress generally increase with the degree of crystallinity, and creep and elongation on breaking are reduced. Strength, fracture toughness and toughness may increase or decrease with crystallinity.

Crystallinity reduces the viscoelastic response and makes polymers non-linear viscoelastic materials.

Crystallinity is favoured by ordered/stereoregular chains. Copolymerisation with a small amount of a second monomer generally eliminates any crystallinity. High melting point crystals are favoured by:

- increased chain stiffness;
- close packing in the crystal;
- strong intra-molecular/interchain forces in the crystal.

Chains can take part in numerous crystallites. A polymer chain can therefore take part in many crystallites and can exist in both amorphous and crystalline domains. In contrast, small molecules are either in the crystalline or amorphous states. Polymer crystals have a disordered amorphous component associated with them and this contributes an unfavourable entropic energy term to the stability of the crystal. Small crystals generally have a greater amorphous component associated with them and generally melt at a lower temperature.

The factors that influence the degree of crystallinity are:

- molar mass and the presence of entanglements. Crystallisation involves disentanglement and chain alignment. Higher molar masses generally restrict diffusion and reduce the degree of crystallinity.
- time and temperature or cooling history. Holding a polymer above its glass transition temperature but below its T_m will provide more time for crystallisation to occur.
- orientation of polymer chains during processing favours crystallisation.

The degree of crystallisation of a polymer may be determined by using X-ray diffraction, differential scanning calorimetry or density measurements.

4.2   Polymer configuration and conformation

The configuration of a polymer is determined by the chemical structure. To convert one configuration to another involves bond breakage. The conformation of a polymer involves its three-dimensional structure, and only bond rotation is required to convert one conformation to another.

A carbon atom with four different groups attached is termed a chiral centre. There are left-handed (L) and right-handed (D) forms of a chiral centre (Fig. 4.3, CD Fig. 4.3). Chiral centres generally result in optical activity. The molecule can rotate plane polarised light. Many polymers have chiral centres.

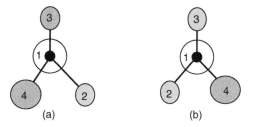

4.3 Schematics of a chiral centre (a) left-handed form, (b) right-handed form (CD Fig. 4.3).

The presence of chiral centres has significant implications for polymers. A polymer with a chiral centre can polymerise in three ways:

- The chiral centres LH and RH can be linked randomly, which is termed atactic.
- The chiral centres can be linked in an alternating fashion, termed syndiotactic.
- The chiral centres can be linked together using one arrangement, which is termed isotactic.

4.3   Tacticity

Vinyl polymers of the type CHX=CHX and CH_2=CXY give rise to syndiotactic, atactic and isotactic forms. Polymers of the type CH_2=CHX are always stereoregular and do not have any tacticity.

The atactic form generally dominates unless special catalysts are used, which serve to orientate the addition of monomers to the growing chain or a template synthesis occurs where monomers are attached to a polymer template prior to polymerisation. Sometimes in a free radical polymerisation the addition of monomers to a growing chain end is not totally random, leading to heterotactic chains. The common form of polymethylmethacrylate is often referred to as atactic but is, in fact, heterosyndiotactic; a mixture of atactic and syndiotactic.

Polymers do not exhibit optical activity if the chiral centre is in the main chain, since, effectively, the groups forming the polymer chain on either side of the chiral centre are practically identical.

4.4   Glass transition temperature

All polymers will undergo a transition from a glassy to a rubbery state at some temperature. This temperature is the glass transition temperature (T_g). It coincides with segmental motion of the polymer chains and a 4–5 order of magnitude reduction in modulus. Diffusion processes increase by several orders of magnitude on passing through T_g. A rubber squash ball is above its T_g and perspex, poly(methylmethacrylate) is below its T_g, at room temperature.

Polymers have time-dependent mechanical properties. They have features of both elastic solids and viscous fluids and their properties, unlike metals and ceramics, change with the time scale of the test. This behaviour is known as 'viscoelasticity'. Polymers are most viscoelastic close to their T_g and exhibit marked changes in moduli, strength, etc. with strain rate close to T_g (CD Fig. 4.4). A detailed discussion of viscoelasticity is not possible here and readers are referred to *An Introduction to Polymers* by Young and Lovell.

4.5   Polymer processing

One of the attractive features of polymers is their ease of processing. Thermoplastic polymers may be moulded to shape by:

- vacuum forming;
- compression moulding;
- extrusion;
- injection moulding.

Vacuum forming is shown schematically in Fig. 4.4. A sheet of polymer is heated with infra-red radiation above the glass transition temperature of the polymer. The polymer is then drawn over a template by means of a negative pressure or vacuum. The templates or moulds are relatively simple to make and relatively inexpensive. However, there are quite severe restrictions on the complexity of the moulding produced.

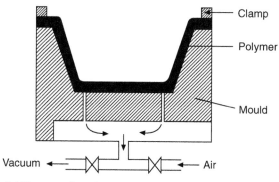

4.4 Vacuum forming.

Compression moulding is shown schematically in Fig. 4.5. Polymer granules are heated in a mould of the required shape. Once the temperature is sufficient and the polymer has become a plastic mass, the two halves of the mould are brought together and the excess polymer is forced between the two halves of the mould. The moulds are relatively simple and cheap to make but the technique is labour intensive and time consuming compared with injection moulding.

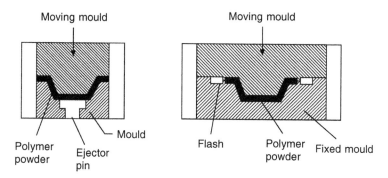

4.5 Compression moulding.

Extrusion is a technique where molten polymer is forced through a die and is used to produce components of a fixed cross-sectional area such as tubes and rods.

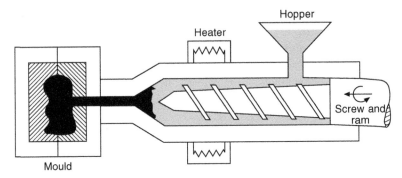

4.6 Injection moulding.

Injection moulding is shown schematically in Fig. 4.6. The technique involves a heated Archimedean screw contained in a barrel, which is fed with solid polymer granules at one end.

4.6   Polymer properties

The properties of thermoplastic polymers are highly dependent on molar mass. Polymers are highly entangled in the melt and solid state. de Gennes proposed a mathematical model for explaining many of the physical properties of polymers based on 'reptation'.

This model (Fig 4.7, CD Fig. 4.5) views a polymer chain as being trapped in a tube of entanglements formed by neighbouring chains. The entanglements restrict motion. Motion occurs by 'wiggling' of the chain along the tube or

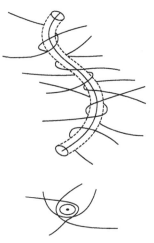

4.7 The reptation model showing one chain trapped in a tube of entanglement formed by neighbouring chains (CD Fig. 4.5).

reptation. de Gennes predicted scaling laws for self-diffusion (D), interdiffusion, dissolution (D_s), and viscosity (η) as a function of chain length.

For example, he predicted; $D \propto MW^{-2}, D_s \propto MW^{-2}, \eta \propto MW^3$

Subsequently, Prentice developed a reptation chain pull-out model for fracture in thermoplastics (Fig. 4.8, CD Fig. 4.6). The model views fracture as occurring by a disentanglement process involving the drawing and stretching of polymer chains across the fracture plane.

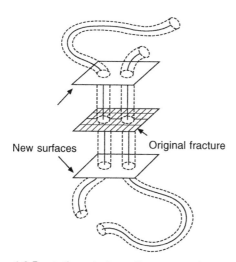

New surfaces   Original fracture

4.8 Reptation chain pull-out model of fracture showing chains being stretched across the fracture plane and disentangled (CD Fig. 4.6).

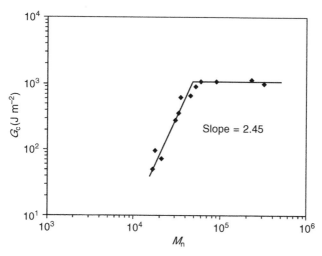

4.9 Log (toughness) plotted against logM_w for poly (methylmethacrylate) (CD Fig. 4.7).

Energy is expended in pulling chains out of their tubes. Prentice's model predicts the toughness (G_C) is proportional to the molecular weight squared. A slope of 2.45 is obtained, rather than the predicted 2.0 (Fig. 4.9, CD Fig. 4.7), which extrapolates to zero toughness at 2.7×10^4 M_n, corresponding to the critical value for chain entanglement. There is a plateau at high molar mass due to chain scission.

4.7   Polymer composites

A composite is a material consisting of two or more phases such as a glass fibre-reinforced plastic (GRP). Most natural structural biological materials are polymeric composites. Bone is a typical example, which is a composite of collagen (a protein) and apatite (a ceramic). By using two phases, we can adjust and manipulate the properties of the composite. Composites may be isotropic, which means the composite has the same properties in all directions, or anisotropic, where the composite has different properties in different directions. Structural biological composites are optimised in terms of their strength to weight ratio. Millions of years of evolutionary development have been invested in their development. Biological materials nearly always have anisotropic properties. Engineers and physicists often regard biological structures as lacking structure and organisation but this is far from reality. Biological composites are often foamed/cellular structures that are composed of an interconnected porous network in order to save weight and allow tissue ingrowth.

4.8   Summary

This chapter introduces the different types of polymers and some applications as biomaterials. The concept of molecular weight is introduced. The molecular weight of the polymer has a large effect on its properties and it is therefore a useful parameter for characterising polymers. Polymers can contain amorphous and crystalline regions but they can never be 100% crystalline. The crystallinity of the polymer greatly affects its mechanical properties, which are time dependent. The chemical structure (configuration and confirmation) also affects the optical properties of the polymer. Polymers have the advantage of being easily processed into complex shapes by methods unique to these materials. Chapter 10 discusses examples of polymers that have been used in biomedical applications.

4.9   Reading list

Billmeyer F.W., *Textbook of Polymer Science,* 3rd edition, New York, Wiley, 1984.

Cowie I.M.G., *Polymers: Chemistry and Physics of Modern Materials*, 2nd edition, London, Blackie and Co., 1993.

Park J.B. and Lake R.S., *Biomaterials – An Introduction*, New York, Plenum, 1990.

Young R.J. and Lovell P.A., *An Introduction to Polymers*, 2nd edition, London, Chapman and Hall, 1991.

5

Biocomposites

IAN D. THOMPSON
Imperial College London, UK

5.1   Introduction

Composite materials are a mixture of two or more phases bonded together so that stress transfer occurs across the phase boundary. Consequently, because stress is not transferred to voids, a porous ceramic, metal or plastic usually is not considered a composite even though the material contains two phases, solid and voids. If a porous structure is infiltrated with tissue, however, it may behave as a composite material, but only if the tissue–material interface is strong enough to transmit stress.

Typically, composite materials are designed to provide a combination of properties that cannot be achieved with a single phase material. A familiar example is fibre glass, in which glass fibres are used to lend stiffness to a polymeric constituent (resin), thus forming a lightweight strong, resilient composite. This concept was employed many years ago in the epoxy–ceramic biocomposite Cerosium®, which was developed by Haeger Pofferies in the USA in an attempt to match the elastic properties of bone. The implants failed, however, because of biodegradation of the polymer and the polymer–ceramic bond.

The failure of Cerosium® is an example of the necessity for a medical implant to meet functional and compatibility requirements simultaneously. Most commercial composite materials, in general, fail to meet this criterion. Table 5.1 summarises the composition and mechanical properties of a number of commercial composites. Elastic properties do not match those of living tissues or there is a lack of biocompatibility for the composite materials. In contrast, bone, which is a composite of collagen and hydroxyapatite (HA), is an ideal composite (Chapter 8). Low elastic modulus collagen fibrils are aligned in bone so that their strong primary bonds are parallel to applied stresses. The high elastic modulus of HA bone mineral provides stiffness and a direct chemical bond between the two phases provides the necessary stress transmission.

Table 5.1 Summary of composition and mechanical properties of some commercial composites

Composition		Volume fraction of fibre or filament	Density (lb/in^2)	Tensile strength (lb/in^2)	Young's elastic modulus (lb/in^2)
Matrix	Fibre or filament				
Epoxy	S –glass	0.60	0.072	290×10^3	7.4×10^6
Epoxy	O	0.40	0.055	103×10^3	22×10^6
Epoxy	Al$_2$O$_3$	0.44	0.063	72×10^3	24×10^6
Epoxy	Al$_2$O$_3$	0.14	0.050	113×10^3	6×10^6
Al	CuAl$_2$	0.45	0.130	42×10^3	10×10^6
Al	Al$_2$Ni	0.10	0.102	48×10^3	11×10^6
Ni	C	0.7		60×10^3	30×10^6
Polymide	B	0.58		130×10^3	37×10^6
Hassalloy X	Mo	0.37	0.323	125×10^3	33×10^6

Figure 5.1 (CD Fig. 5.8) shows the mechanical properties of new biocomposites that satisfy both functional and biocompatibility requirements compared with other structural materials. With such materials there is the potential to produce a lightweight and high strength device with anisotropic properties similar to natural bone. This can be accomplished by orienting the high modulus phase in specific orientations. Two alternative orientations, representing extrema, are shown in Fig. 5.2. The ratio of the second-phase modulus (E_2) to the matrix modulus (E_1) and volume fraction of the second phase particles significantly affects the elastic modulus of the composite (Fig. 5.2). By controlling these variables, it should be possible to achieve

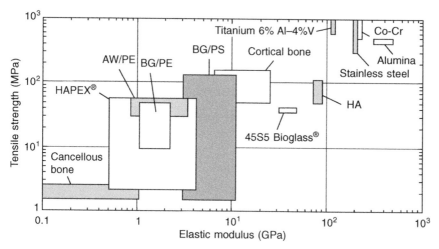

5.1 Mechanical properties of various biomaterials (Thompson and Hench, 1998) (CD Fig. 5.8).

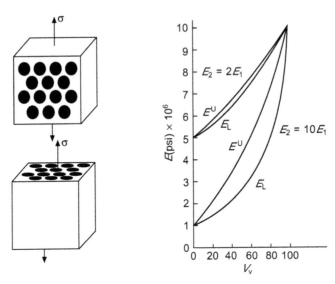

5.2 Schematic of the effect of orientation on the elastic modulus of materials.

nearly any desired mechanical behaviour for a composite and thereby emulate the properties of bone.

Applications of composites in the physiological environment are often limited by the stability of the interphase bonds. Should breakdown of the interphase bonds occur, rapid mechanical deterioration follows and degradation products may be physiologically hazardous. Because of the potential of composite systems, considerable research on the subject is still in progress. The following sections of this chapter review the status of development and application of various types of biocomposites. Details of alternative biocomposites are given in the CD lecture for Chapter 5.

5.2   Bioactive ceramic polymer composites

The most successful biocomposites were designed by Professor W. Bonfield's research team at the University of London to match the natural components of bone: 45vol.% hydroxyapatite and 55vol.% collagen. For the composite, natural HA was replaced with particulate HA produced by a synthetic route. The collagen was replaced by polyethylene (PE). The resultant material was designed as a bone analogue rather than as a load-bearing orthopaedic device and it was named HAPEX®.

All new biomaterials are developed/designed to fulfil a specific clinical need. The HAPEX® material has succeeded because clinicians and material scientists were jointly involved in the research. The development of this important composite has been in five stages and provides a good example of technology transfer (Chapter 24) and regulatory issues in biomaterials (Chapter 23).

Pilot studies (1979–81)

A range of composites of HA particulate and high density polyethylene were made with HA volume fractions ranging from 0.1 to 0.6. The resultant Young's modulus ranged from ~1 to ~8 GPa and strain to failure from >90% to 3% depending on the volume fraction of HA (Fig. 5.3).

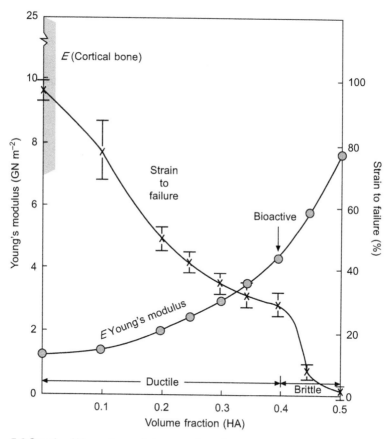

5.3 Graph of Young's modulus as a function of volume fraction of HA in HAPEX®.

Laboratory testing (1982–88)

Repeated mechanical testing of the composite HA/PE showed that the upper limit of HA addition was a volume fraction of 0.4. The resultant fracture toughness was comparable to that of cortical bone. However, *in vivo* biological testing of the composite showed that HA volume fractions of less than 0.2 caused the material to become biologically inert. A non-adherent fibrous capsule covered the composite and no adherent bonds were formed between the implant and the host tissues. Hence a range of 0.2 to 0.4 volume fraction of HA was determined to be optimal. Controlling particle size and morphology in important tests showed that, with a D_{50} (the equivalent spherical diameter at a cumulative percentage of 50%) particle size of 4 µm and a convoluted morphology, strong filler/matrix coupling was achieved by mechanical processing alone. A chemical coupling method has also been developed but is not used clinically at present.

Clinical trials and pilot plant production (1989–93)

The non-load-bearing HA–PE bone analogue material was evaluated clinically in a suborbital floor reconstruction. The trial identified that the composite could be trimmed by the surgeon with simple surgical tools whilst in theatre. This allowed the surgeon to obtain a better fit, and the implant to have more contact with the host's tissues. Fifty patients received implants and follow-up showed that the trial had a good outcome. The material was produced by a twin screw co-rotating extruder, which produced a homogeneous distribution of the HA phase in the polyethylene matrix. The success of this first clinical trial was followed by establishing a dedicated clean processing facility at the University of London IRC in biomedical materials.

Licensing (1994–95)

The material was licensed to a commercial manufacturer as a bone substitute with primary applications in ear, nose and throat (ENT) surgery. The US Food and Drug Administration subsequently gave the material regulatory approval allowing the British technology to be sold in the USA and a CE Mark was granted (see Chapter 23).

Commercialisation (1995)

The HA–PE composite was given the trade name HAPEX™ and worldwide clinical use followed.

The original idea of using HAPEX™ as an orbital floor reconstruction material was expanded to include its development into a middle ear prosthesis. Hearing is transferred from the tympanic membrane, to the eardrum, to the inner ear by three bones, which conduct and amplify the vibrations of the eardrum: the hammer, stirrup and anvil. Conductive hearing loss is a common problem and is due to the inability of the obsicular chain to conduct acoustic vibrations to the cochlea. Hence, one, two or all three of the bones are removed by an ENT surgeon and replaced by a middle ear device. Standard materials used for these devices have been PTFE and other polymer-based devices. A monolithic Bioglass® device had been developed and found to be a clinical success but is difficult to shape in the operating theatre. This left a clinical market open for the HAPEX™-like device.

5.3   Design criteria for biocomposites

The bioactive ceramic composites section described a successful bone replacement material for middle ear prostheses. However, there is no scale on which to determine the best bioactive composite for load-bearing bone analogue materials.

A quality index has been proposed to evaluate the quality of a material when compared to that of natural bone. The index is based on the fact that the following properties are important when evaluating artificial bone replacement materials:

1. elastic modulus;
2. strength;
3. fracture toughness;
4. bioactivity.

Using the four variables, it is possible to produce a quality index (I_q) of a material's performance compared to natural bone:

Quality index (I_q) = (fracture toughness × index of bioactivity × tensile strength)/Young's modulus

This order of properties is an attempt to evaluate materials that have a relatively high strength and a moderate to low elastic modulus to avoid stress shielding of bone. The required properties for an I_q evaluation are shown in Table 5.2. Cortical bone has an I_q value of 500, whilst cancellous bone has a value of only 8. The fracture toughness component of the equation is possibly given more influence over the I_q value than is required in reality. However, fracture toughness has been established by Bonfield as an important property when evaluating bone substitutes. It should be noted that values assigned to each material are approximations; more work is required to evaluate fracture toughness of bioactive composites. The results indicate that Bioglass® polysulphone materials are better at replacing cortical bone whilst Bioglass® alone is better at replacing cancellous bone. The interface between the polysulphone and Bioglass® is of primary concern in raising the I_q to that of the cortical bone. The modified polysulphone is the most likely material to reach the target values. This ranking identifies that improvement of fracture toughness for the BG/PS composite is a potential research path to follow in

Table 5.2 Calculation of quality index of some biomedical composites. K_{1C} = fracture toughness, E = Young's modulus, UTS = ultimate tensile strength. Note: All values are median values from a range of mechanical properties data. Estimated lines are underlined

Ceramic phase	Polymeric phase	K_{1C}	E (GPa)	I_b value	UTS (MPa)	Quality index $I_q = \dfrac{(K_{1C} \times I_b \times UTS)}{E}$	$\dfrac{I_q \text{ (material)}}{I_q \text{ (cortical)}}$	$\dfrac{I_q \text{ (material)}}{I_q \text{ (cancellous)}}$
Cortical		6.0	15	13	100	500	1.00	1.0
Cancellous		0.1	1	13	3	8	0.02	1.0
HA		1.0	85	3	80	3	0.01	0.4
Bioglass		0.6	35	13	42	9	0.02	1.2
A/W GC		2.0	118	6	215	20	0.04	2.7
BG-C 40%		1.0	68	13	210	37	0.07	5.0
BG-C 100%		0.8	80	13	200	24	0.05	3.2
HA (0.4V_f)	PE	3.0	4	3	23	49	0.10	6.5
Bioglass (0.4V_f)	PE	3.0	3	13	10	150	0.30	20.0
Bioglass (0.4V_f)	PS	3.0	7	13	52	291	0.58	38.8
Bioglass (0.4V_f) (modified)	PS	3.0	5	13	103	757	1.51	101.0

order to match cortical bone. Adjusting the K_{1C} to 2.00 MPa m$^{1/2}$ will produce an I_q value equal to that of bone. Rounding of particles and controlling size distribution and interface bonding could make such an improvement possible.

The I_q data suggest that Bioglass®/polysulphone composite material has properties closest to those of cortical bone. This material also has a class A bioactivity and a very high I_B value. This combination of properties suggests that it may be a valuable material in orthopaedic applications. However, as so much data is still unpublished, it is not possible to calculate I_q values for all composites.

5.4   Inert ceramic composites

The use of carbon and inert glass fibres has also been widely explored. These standard engineering fibres have been used mainly in reinforcement of orthopaedic devices, such as femoral hip stems, knee prosthesis and fracture fixation plates. Because of the inability of carbon and inert glass to form any bioactive bonds, they have not been used in bone analogue materials.

A study by Jockisch, in 1992, showed that carbon fibre-reinforced poly-ether-ether-ketone (PEEK) has good mechanical properties. The fibrous capsule thickness around carbon-reinforced PEEK was smaller than unreinforced ultra high molecular weight polyethylene, indicating less micromovement of the carbon/PEEK device. Toxicology screening showed the device to have some debris present but this did not cause any major foreign body reaction. This type of fracture fixation plate has been used clinically but generally found not to be as reliable or biocompatible as metallic plates.

Twenty years of clinical experience show that metallic hip stems stress shield the femur. This has led to research into uses of composites in total hip replacements. One example is carbon fibre reinforcement of polysulphone. This design had a core of unidirectional carbon/polysulphone with bidirectional braided outer layers. The surface of the stem was covered with pure polysulphone, possibly to aid in the press fitting of the device into the femoral shaft. Finite elemental analysis (FEA) of the design showed that it would produce minimal disruption to the physiology of the host cortical bone. Implantation of the device and follow-up after 4 years showed that the canine host had a favourable remodelling response. Carbon fibre reinforcement of epoxy-resin has also been proposed as a potential material for hip stems. However, FEA of the material has shown up concerns for load transfer from the femoral neck, which results in this area having a high risk of fracture.

Other areas of total hip replacement have been targeted for composite technologies. Acetabular cups are traditionally made of polyethylene, but have been shown to cause osteolysis (degradation of the host cortical bone) due to wear debris. Efforts have been made to improve the durability of such devices by adding carbon fibre reinforcement. The use of carbon fibre-reinforced plastic acetabular cups was tested by articulating them against alumina femoral heads in a hip simulator. The test was conducted at 2500 N loads with an articulation frequency of 0.857 Hz. The femoral head wore the cup sufficiently to allow the head to move into the cup by 10 μm after 500 000 cycles. After 1 million cycles, the additional change was only 1 μm. It was concluded that the cup was suitable for implantation, considering that retrieved cups wear by 6 μm each year. However, other experimental results have shown that these devices do not respond well when implanted and perform less well than the original material. As the device breaks down, carbon is released into the host tissues, which may cause cysts, tissue inflammation and other toxicological responses.

Composites have also been examined for use in knee prostheses. The two components of the knee prostheses articulate together like a natural knee joint, but in the natural joint synovial fluid and cartilage reduce the friction of the articulating surfaces. In order to achieve low friction articulation, a polymeric surface is introduced onto the tibial plate of the prostheses. In elderly patients the normal polymeric tibial component, ultra high molecular weight polyethylene (UHMWPE) is adequate, but for younger patients a longer implant lifetime is required. The tibial component is susceptible to creep, hence, carbon fibre reinforcement of the device has been considered. Mechanical testing of such devices has shown that properties can be increased by a factor of two. However, tissue inflammation, due to the presence of carbon fibres in the articulating surfaces has also been seen in such knee prosthesis.

Composite materials have a potential use in load-bearing orthopaedic applications, but very few have been tested clinically. Composites used for articulating surfaces have a tendency to cause inflammatory problems due to the loss of carbon particles.

5.5   Resorbable polymer matrices

All of the materials detailed in the previous sections have been used with resorbable α-polyester matrices. The two principal polymers used are poly(glycolic acid), PGA, poly(lactic acid), PLA, and co-polymers of the two (see Chapter 10). Polylactic acid degrades to produce non-toxic lactic acid, which is metabolised to carbon dioxide and water, and is easily excreted. The advantage of these polymers is that they have time-varying mechanical properties; resorption rate is reduced by increasing the volume fraction of PGA. Additions of randomly orientated chopped carbon fibre have shown to have improved mechanical properties to that of the pure polymer, but strengths are still not sufficient for load-bearing applications. Long carbon fibre reinforcement with 0° and +/–45° orientation was tried. Initial properties were adequate but the fibres did not bond to the PLA and delamination resulted in rapid loss of strength. Hence, fibre polymer coupling was required. A totally degradable implant was developed using calcium phosphate glass fibres. Mechanical properties were promising but the degradation rate of the composite resulted in failure of the prosthesis before the bone fracture was able to heal.

A segmented copolymer of poly(polyethylene oxide terephthalate) PEOT, and poly(butylene terephthalate) PBT, has been used as a resorbable bone analogue, Polyactive® (PA). Tests of this material in a rabbit defect model suggest that PA allows more bone ingrowth than an empty control site, but that PA does not stimulate bone ingrowth in the same fashion as bioactive ceramics. Self-reinforcing composites formed from PLA and PGA have also been mechanically tested along with additional carbon fibre reinforcement.

These polymers and their composites are not suitable for major load-bearing applications, but are suitable for small non-load-bearing fracture fixation devices, i.e. finger joint replacement. However, the interest in this region of biocomposites is very high indeed. There are two obvious and distinct advantages of these types of materials: (i) as the polymer resorbs there is less stress shielding, (ii) the implant simply dissolves away so there is no need for a second operation, unlike with the metal plates used in fracture fixation.

Further development of processing techniques for these composites has resulted in bioactive ceramic phases to be added. Additions of β-tricalcium phosphate (TCP) to copoly-L-lactide (CPLA) has resulted in a composite with a bending strength of 51 ± 6 MPa, Young's modulus of 5 ± 1 GPa and a fracture strength of 52 ± 3 MPa. Thus, this material is approximately half the required properties of cortical bone, but does have the advantage of being osteoconductive and the matrix is resorbable. Additionally, apatite materials have added to porous poly-L-lactic acid. This material has shown similar *in vitro* potential and mechanical problems as the TCP/CPLA material. It is this class of biocomposites that is likely to take on an increased role in the future.

5.6   Conclusions

It has been shown repeatedly that composites have an important role in modern medicine, but the development of such technology is time consuming and expensive. Many surgeons demand an immediate solution from the materials scientist, which can result in a premature clinical trial that has not undergone a series of mechanical and toxicological tests. The incidence of composites becoming serious problems in the clinic are very small; however, one noteworthy material is Proplast®. This material was developed as a composite of poly(terafluoroethylene) and pyrolysed carbon, for use in soft tissue augmentation, where it was clinically successful. The resultant black material had 70–90% porosity with pores ranging from 80–400 μm, which facilitated ingrowth of soft connective tissues. Use of this composite in load-bearing applications, such as the temporal mandibular joint (TMJ) led to failures. When implanted on chin sites in canines, the material showed very poor biocompatibility. All implants were loose and had no bone ingrowth into them. However, there were large amounts of fibrous tissue, which contained giant cells, macrophages and leucocytes. There was also some evidence of the material eroding the underlying bone. The adjacent tissues also contained granulomas with contained fragments of Proplast® material. The major failing of this material is the breakdown and release of carbon, which moved out of the material and set up an inflammatory response. Testing of this material in load-bearing applications should have highlighted the potential problems and avoided the clinical failures.

5.7   Summary

This chapter introduces the idea of composites and describes the principle of combining the strength of a ceramic with the toughness of a polymer to match the properties of natural tissues. Examples of bioinert and bioactive composites used in biomedical applications are given, focussing on HAPEX®, which has arguably been the most successful commercial composite for clinical use. Their mechanical properties are compared with bioceramics and natural tissues. The potential and high interest of bioresorbable scaffolds is discussed.

Many composites tested have failed to achieve clinical success because a new technological development has often made the product outdated and no longer the best solution to the clinical need.

5.8   Reading list

Bronozino J.D., *The Biomedical Engineering Handbook*, Boca Raton, Florida, CRC Press, 1995.

Hench L.L. and Ethridge E.C., *Biomaterials, An Interfacial Approach*, New York, Academic Press, 1982.

Hench L.L. and Wilson J. eds., *Clinical Performance of Skeletal Prostheses*, London, Chapman and Hall, 1996.

Thompson I.D. and Hench L.L., 'Medical applications of composites', *Comprehensive Composite Materials*, 2000, **6** (39), 727–753.

Totora G.J. and Reynolds Grabowski S., *Principles of Anatomy and Physiology* 10th edition, New York, Wiley, 2003.

Tsuruta T. and Nakajima A., *Multiphase Biomedical Materials*, the Netherlands, VSP Utrecht, 1989.

6
Cells and tissues

JUNE WILSON HENCH
Imperial College London, UK

6.1   Introduction

Understanding the performance of biomaterials in the body requires collaborative research among many disparate disciplines. In its simplest form the effect of the physiological environment on materials is determined by materials scientists and engineers, and the effect of the material on the body by biologists and clinical scientists. To be productive, it is essential that each member of a multidisciplinary team must understand the terminology and basic principles of the others so that they may communicate ideas, understand the literature and contribute to the work as a whole.

In Chapters 2–5 the relevant principles and techniques of materials science and engineering are described for biologists and clinicians. In this chapter the terminology and descriptions of the tissues and their physiological environment are described to facilitate an understanding of the biological principles involved in implantation of biomaterials.

You will be shown the four tissue groups, how they combine to form organs and how they appear in histological sections. All the diagrams and photomicrographs here and in the accompanying part of the CD are of tissues as seen under the light microscope.

6.2   Definitions

Epithelium

The tissue that covers, lines and protects organs. It can absorb, secrete and excrete a variety of substances.

Connective tissue

The tissue that wraps, connects, nourishes and supports all other tissues and organs. Blood, soft tissue, bone, fat and cartilage are all connective tissues.

Muscle

The tissue consisting of fibres which, acting together, contract and relax to produce movement. Muscle may be under voluntary control (skeletal muscle) or involuntary control (the heart and the muscle in the lungs and gut).

Nervous tissue

The tissue that transmits and conducts signals to and from organs. It may be central (brain and spinal cord) or peripheral (nerves).

Organ

A functioning unit within the body made up of various combinations of tissues.

Histology

The light microscopical appearance of normal tissues and organs.

Histopathology

The light microscopical appearance of abnormal tissues and organs.

Toxicity

An adverse effect producing changes (pathology) that can be detected by histopathology.

Biocompatibility

The possession of both of two essential properties; the lack of toxicity and effective function. (This may be the most misused term in biomaterials research when used to describe a material. Biocompatibility is not a property of a material.)

Cilia

Small hair-like projections from the free surface of a cell, which move material across cells by synchronous beating.

Exocrine gland

One that passes its product through a duct.

Endocrine gland

One that passes its product directly into the bloodstream.

Macrophage

A cell of around 10–12 μm in size that can ingest and digest or excrete particles of foreign material.

Microphage

A cell of 3–5 μm in size, which can ingest and digest or excrete foreign macromolecules and very small particulates.

Osteoblast

A cell that produces bone.

Osteoclast

A cell that removes bone.

Osteocyte

A resting bone cell.

Fibroblast

A cell that produces fibres.

Chondroblast

A cell that produces cartilage.

Chondrocyte

A resting cartilage cell.

There are only four basic tissue groups – epithelium, connective tissue, muscle and nervous tissue – and, together, they form organs. Although all organs contain all four component tissues, connective tissues are the most important in this field. All experimental procedures and applications of biomaterials must breach connective tissues and it is the response of those tissues, which controls the success or failure of the device.

6.3   Epithelium

Epithelial cells cover, line, absorb and excrete. They are described by their appearance in histological sections.

1. They may be simple, occurring in a single layer.
2. They may be complex, in multiple layers.

Whether simple or complex, three shapes for the individual cells can be distinguished.

Squame cells

A squame is an individual flattened cell in which the central nucleus protrudes (Fig. 6.1, CD Fig. 6.2). These cells form squamous epithelium. A simple example of squamous epithelium is the endothelium that lines blood vessels (Fig. 6.2, CD Fig. 6.3). Where several layers of squamous cells are seen, they form a complex epithelium. The layers will be in one of two arrangements.

6.1 Schematic of a squame cell (CD Fig. 6.2).

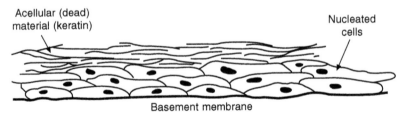

Acellular (dead) material (keratin)

Nucleated cells

Basement membrane

6.2 Schematic of a squamous epithelium (CD Fig. 6.3).

In the first arrangement there are many layers, those at the base having distinct nuclei that are absent from the superficial layers, which consist of acellular (dead) material. This is typical of a protective epithelium, which replaces itself from the base when subjected to damage or abrasion. The dead material is keratin, which is a protein. Such epithelium is found in the skin and upper and lower ends of the gut, the upper end of the respiratory tract and the vagina.

The second form of complex stratified epithelium appears as layers with nucleated cells throughout. It is known as transitional epithelium (Fig. 6.3, CD Fig. 6.5) and occurs in the urinary tract, notably lining the bladder. This epithelium can conform to the shape of the bladder, whether it is full or empty, providing a continuous layer of lining cells, when distended, to protect

Basement membane

6.3 Schematic of a transitional epithelium (CD Fig. 6.5).

underlying tissue from the damaging potential of urine and allowing the same cells to pile up in layers when the bladder is empty.

Cubical cells

Cells which appear to have similar height and width in cross-section are termed cubical (sometimes cuboidal). These cells sit on a membrane at their base, which is essential to their function. The simple epithelia that cover and line organs such as the kidney and ovary are cubical cells arranged in a single layer (Fig. 6.4, CD Fig. 6.8).

Basement membrane

6.4 Schematic of cubical cells (CD Fig. 6.8).

Cubical cells do not appear in multiple layers, but can be termed complex when they are modified by being ciliated. The presence of fine hair-like threads on the free face of the cell shows that the cell is capable of moving particulate or similar material by synchronous beating of these hairs. Such cells may be found in the Fallopian tube where they move the ovum (Fig. 6.5, CD Fig. 6.10).

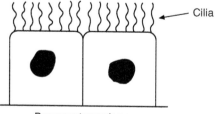

Cilia

Basement membrane

6.5 Schematic of ciliated epithelia (CD Fig. 6.10).

Columnar cells

Cells in which the height is several times that of the width are termed columnar. These cells are found in much of the respiratory system and most of the gut (Fig. 6.6, CD Fig. 6.11).

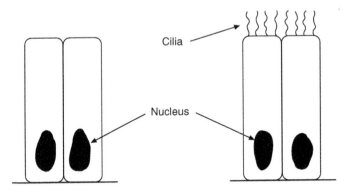

6.6 Schematic of columnar cells (CD Fig. 6.11).

Columnar cells may also be ciliated to move particulate or similar matter. Ciliated columnar cells are found in the respiratory tract, including the middle ear, and in the testes where they move the sperm before the sperm become motile. Ciliated cells do not occur in the gut; the cilia are too fragile to withstand the passing of the gut contents.

There is no true stratification of columnar epithelium, but in the gut pseudo-stratification is seen (Fig. 6.7, CD Fig. 6.13). What appears to be a piled up layer of columnar cells with nuclei in different positions is, on close inspection, a crowded layer of cells in which the shape of the cell may be distorted but can always be shown (by special staining if needed) to have a contact with the basement membrane. The reason for this appearance is the same as that for transitional epithelium; it allows the lining of the gut to compensate for the relatively large changes in shape and dimension during digestion and excretion.

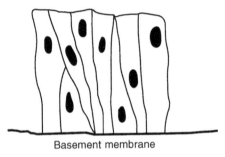

Basement membrane

6.7 Pseudo-stratified columnar epithelium (CD Fig. 6.13).

The columnar epithelia of the digestive and respiratory systems may be further modified to secrete mucus. In the digestive system, mucus produced by these cells protects the stomach lining from acid digestive juice and later lubricates the passage of faeces as they become dehydrated in the lower part of the gut (Fig. 6.8, CD Fig. 6.15). In the respiratory tract the mucus serves to trap bacteria and particulates present in the lungs and bronchi. Such mucus is then removed by coughing or swallowing.

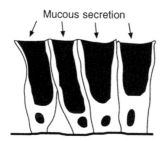

6.8 Mucous columnar epithelium (CD Fig. 6.15).

Glands

The secretion of materials into the various systems of the body, including the bloodstream, is the property of certain cubical and columnar epithelia. The cells form structures known as glands. The secretions may be digestive enzymes, as in salivary glands; more complex collections of substances, as in the liver or mammary gland; or hormones as in the adrenal, pituitary and similar organs. The collected cells organise into groups with a central duct to carry the secretion away. These are known as exocrine glands (Fig. 6.9, CD Fig. 6.19). Salivary, mammary, liver, sweat and pancreas are examples of exocrine glands.

Groups of glandular cells with no duct system, which secrete directly into the bloodstream, are known as endocrine glands. Adrenal, pituitary, thyroid and the islet cells of the pancreas are examples of endocrine glands.

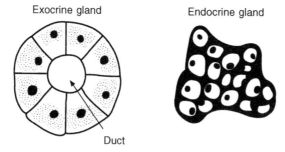

6.9 Schematics of an exocrine gland and an endocrine gland (CD Fig. 6.19).

6.4   Connective tissue

There are many types of connective tissue with apparently little in common other than their presence in and around the organs. All have three components; cells, fibres and ground substance. A distinction is often made between hard tissue (bone, enamel and dentin) and soft tissue. It is important to remember that there are several types of soft tissue with properties ranging from fluid (blood) to flexible (dermis and fat) or stiff (cartilage). It is variation in the ground substance and fibres that allows these seemingly diverse tissues to form a group.

Connective tissues contain two classes of cells: those that produce the fibres and ground substance and those cells of the body's defence systems that travel within it (Fig. 6.10, CD Fig. 6.21). The latter are the cells that ingest and remove foreign material such as bacteria, tissue fragments and material fragments. They are called microphages or macrophages depending on the size of the particle that can be ingested. They are part of the inflammatory process and are discussed in more detail in the following chapter.

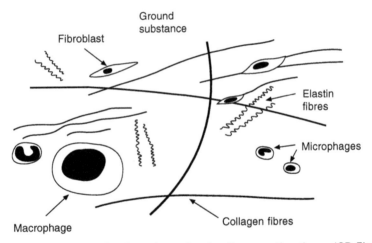

6.10 Schematic of cells and matrix of soft connective tissue (CD Fig. 6.21).

The fibres of connective tissue provide the mechanical properties of the tissue and are mainly of collagen or elastin. The ground substance in which these components live and move provides support and a means of transport for nutrients and products of excretion.

Depending on the relative proportions of the three components, connective tissue may be described as loose, dense, fibrous or elastic. The cells in bone include osteoblasts, osteoclasts and osteocytes, responsible for the laying down and contouring of bone. In the dentin of teeth the process is essentially the same. The fibres are principally collagen fibres that are mineralised to give strength. Bone is dealt with in more detail in Chapters 8, 12 and 13. In the dermis (dense or fibrous connective tissue) the cells include fibroblasts, which produce collagen and elastin, and macrophages that interact to control the production of scar tissue, as well as normal tissue. In adipose or wrapping connective (loose or areolar connective) tissue, the cells include fat cells and there are relatively few fibres. The cartilage of the pinna of the ear and tip of the nose (elastic cartilage) contains chondroblasts as the cellular component and a predominance of elastin fibres to give flexibility. In this, as in other types of cartilage, the ground substance is dense and without an efficient blood supply. See Chapter 8 for details.

All nutrients and products of excretion move through the ground substance by diffusion, which is an inefficient means of distribution and which must be taken into account when cartilage is studied. Blood, which is a special case, does conform to the pattern of connective tissues. It has a range of cells, both red cells and white. The fibres are soluble and are responsible for clotting when precipitated and the ground substance is the serum, the liquid part of blood without the plasma proteins. Blood will be discussed in more detail in later chapters.

Bone, tendons, ligaments, cartilage

These special connective tissues are described in detail in Chapter 8.

6.5   Muscle

Muscle is made up of bundles of individual fibres, each with the ability to contract and relax. This property derives from the contractile proteins that make up the fibres. Muscles move the body via bones and joints and control the movement of the gut, the respiration and the circulation. There are three main types of muscle.

1. Smooth muscle appears under the microscope as bundles of strap-shaped cells, each with a centrally placed nucleus, which show no particular staining characteristic (Fig. 6.11, CD Fig. 6.27). Smooth muscle is found throughout the respiratory system, the digestive system and blood vessels. It cannot be controlled consciously and is also termed involuntary muscle.
2. Skeletal muscle is seen as bundles of fibres in which the individual cells cannot be distinguished, the nuclei are at the periphery of the bundle and

6.11 Schematic of smooth muscle cells (CD Fig. 6.27).

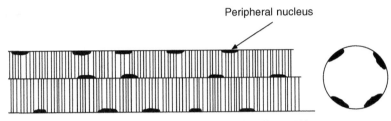

6.12 Schematic of voluntary muscle cells (CD Fig. 6.29).

the tissue is clearly striped (Fig. 6.12, CD Fig. 6.29). The presence of the striping can be enhanced by special staining. It is under voluntary control and is also termed voluntary muscle.

3. Cardiac muscle is found only in the heart and appears as bundles of striped muscle fibres with centrally placed nuclei (Fig. 6.13, CD Fig. 6.33). Branching of the bundles can be seen and under the right conditions structures called intercalated discs, which occur only in this muscle, can be found. Cardiac muscle is an involuntary muscle.

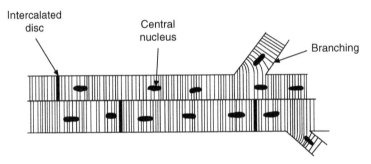

6.13 Schematic of cardiac (involuntary) muscle cells (CD Fig. 6.33).

6.6   Nervous tissue

Nervous tissue occurs centrally, in the brain and the spinal cord and peripherally as the bundles of fibres that carry messages to all parts, which are commonly called nerves (Fig. 6.14, CD Fig. 6.35).

Central nervous tissue is found in brain and cord and is made up of cells called neurons, which communicate with each other by a network of fine processes called dendrites and which pass messages over greater distances via a single long process called an axon. The axon is insulated by a waxy discontinuous coating called the myelin sheath. To the naked eye the areas that are mainly cells appear grey and the coated axons appear white. This is the origin of the terms grey and white matter still used to describe the central nervous system.

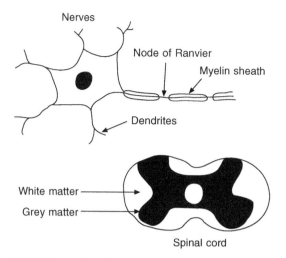

6.14 Schematic of a nerve cell and a cross-section of the spinal cord (CD Fig. 6.35).

The nerves of the peripheral nervous system are essentially bundles of these axons that run with attendant connective tissues to all parts of the body.

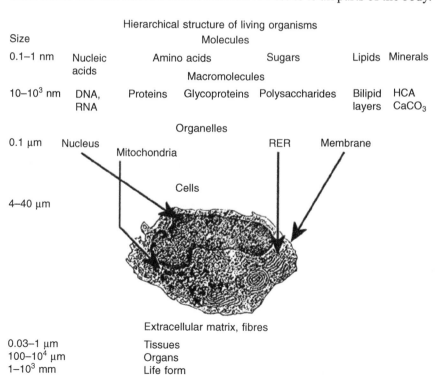

Hierarchical structure of living organisms

Size					
		Molecules			
0.1–1 nm	Nucleic acids	Amino acids	Sugars	Lipids	Minerals
		Macromolecules			
10–10³ nm	DNA, RNA	Proteins Glycoproteins Polysaccharides		Bilipid layers	HCA CaCO₃
		Organelles			
0.1 µm	Nucleus	Mitochondria	RER	Membrane	
		Cells			
4–40 µm					
		Extracellular matrix, fibres			
0.03–1 µm		Tissues			
100–10⁴ µm		Organs			
1–10³ mm		Life form			

6.15 The hierarchical structure of living organisms as a function of size (CD Fig. 1.2, 6.1).

6.7   Summary

There are four tissue groups, epithelium, connective tissue, muscle and nervous tissue. They combine together in various proportions to form organs in a hierarchical structure. Figure 6.15 (CD Fig. 1.2/6.1) summarises how this structure is built up.

6.8   Reading list

Alberts B., Bray D., Lewis J., Raff M., Roberts K. and Watson J.D., *Molecular Biology of the Cell*, 3rd edition, New York, Garland Publishing, Inc., 1994.

Bloom W. and Fawcett D.W., *A Textbook of Histology*, 10th edition, Philadelphia, W.B. Saunders Co., 1975.

Karp G., *Cell and Molecular Biology: Concepts and Experiments*, New York, J. Wiley & Sons, Inc., 1996.

Pollack G.H., *Cells, Gels and the Engines of Life*, Seattle, L. Ebmer and Sons, 2001.

Ratner B.D., Hoffman A.S., Schoen F.J. and Lemons J.E., *Biomaterials Science: An Introduction to Materials in Medicine,* 2nd edition, New York, Academic Press, 2004.

7
Inflammation and wound healing

J U N E   W I L S O N   H E N C H
Imperial College London, UK

7.1   Introduction

The aim of this chapter is to create an understanding of the reactions during the process of normal wound healing, the rate at which they occur and the changes to those reactions that are the result of the presence of an implanted biomaterial. These changes may be due to various properties of the implant: its shape, size, chemistry and mechanical properties of the surface or of the bulk. The relevance of these changes is a critical part of the study of the implant and its material.

7.2   Definitions

Inflammation

The condition produced by tissue damage, characterised by redness, warmth, tension and pain.

Phagocyte

Any cell that can ingest foreign material.

Microphage

Small cells (3–5 µm) that engulf small particles such as bacteria, macromolecules or tissue fragments. They are more commonly called polymorphs.

Polymorphs

Also known as PMNs, these are circulating white cells or leucocytes that move from the blood vessels into the connective tissue when required.

Macrophages

Larger cells (10–12 μm) that can engulf, digest and transport larger particles than can a polymorph. The macrophage derives from a different leucocyte (a monocyte), which moves from blood vessels into the tissues, first in a resting state, where it is termed a histiocyte and then, when it is needed as an activated macrophage, it moves to the damaged tissue.

Giant cells

Very large cells (40–100 μm), which form in connective tissue when the material to be removed is too large for individual macrophages. They are formed when many macrophages are crowded together and undergo mutual dissolution of their cell walls to form a large syncytium of cells. Giant cells may have several hundred nuclei and are often seen on a rough surface in a futile attempt to digest surface irregularities. Giant cells have a relatively short life and cannot reproduce, so their presence some time after implantation is a sign of an unwanted, persistent effect.

7.3   Effects of implantation

The implantation of any biomaterial causes damage to the host tissue and inevitable inflammation. The consequences of tissue damage during surgery must be distinguished from those due to the implant. Persistent tissue changes, which are due to the presence of the material, may be produced either mechanically or chemically. These changes must be understood so that they may be controlled and manipulated for a particular application. Wound healing and inflammation are controlled by effects in the connective tissue, which carries the blood supply. Without an adequate blood supply there is no wound healing.

To understand the way in which tissues respond to an implant, it is first necessary to understand the normal wound healing process. In this chapter the normal process of inflammation and wound healing is described and the ways in which the properties of the implant can affect that process are indicated.

7.4   Normal wound healing

In wound healing there are three stages which are conveniently recognised as:

1. the cellular phase in which there are vastly increased numbers of cells over those seen in normal tissue. Initially these are mostly phagocytes, but as the reaction proceeds the phagocytes diminish and fibroblasts multiply;

2. the fibrous phase during which fibres, notably collagen, are produced by the fibroblasts;
3. the resolution or maturation phase during which the collagen fibres are aligned and remodelled to conform as closely as possible to the original tissue.

Inflammation is a local effect seen as reddening, warmth and tension in the tissues. Damaged tissue and bacteria release chemicals, mainly histamines and prostaglandins, which cause the blood vessels to dilate to bring in more blood and thus more white cells. This produces the characteristic signs of inflammation. The dilatation of the vessels becomes enough to allow leakage of fluid and cells into the connective tissue (Fig. 7.1).

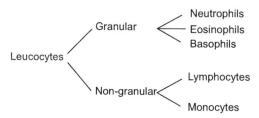

7.1 Classification of white blood cells.

The first wave of white blood cells are the microphages (or polymorphs), which can phagocytose bacteria and small molecules. They can digest these materials once ingested. The tissue macrophages (histiocytes) next become mobile and are joined by an influx of monocytes from the blood to replace and augment them. The fluid that leaks into the tissues causes the swelling and tension associated with inflammation. It dilutes noxious material and facilitates the movement of cells. These cellular events are the first step in wound healing. Any breach is filled by clotted blood, which restores continuity and provides a scaffold for cells to move to the damaged area along concentration gradients of chemicals. An important factor is that polymorphs with ingested material die at the site, so that any wound can contain dead cells as well as what attracted them in the first place. If this was infection, then the accumulation of bacteria and dead polymorphs form pus, which must be removed before the healing process can proceed.

However, in most situations infection and pus are not present and this early phase peaks around 4 hours after damage. Following the polymorphs, the macrophages move into the area to continue the removal of damaged material. This stage peaks by about day 3. While these cells are active, they produce enzymes, which inhibit the production of collagen fibres by fibroblasts. As the wound is cleaned by the macrophages, new blood vessels begin to invade the area. Fibroblasts proliferate where the partial pressure of oxygen is high, adjacent to new blood vessels and increase in numbers to reach a peak by day 5.

By day 7, the cells are mostly fibroblasts and the remaining macrophages become inactive. There is a fine balance between macrophage and fibroblast so that proliferation of fibroblasts and fibrogenesis occurs at the appropriate time. When fibroblasts are released from the inhibition of collagen production, they lay down collagen in a random fashion. This takes up to day 12.

The final stage, during which strength develops, comes next. Sometimes called the maturation phase, this is when the collagen fibres organise and become aligned along the lines of stress in the area. By the combined actions of collagenases, enzymes which break down collagen and fibroblasts which produce it, the final amount and orientation of collagen at the site approaches that of the normal tissue. This is the ideal result in a healed wound. It will never, however, be as strong as the normal tissue. These time lines are shown in Fig. 7.2.

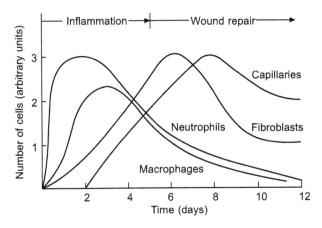

7.2 Time dependence of cellular changes in wound healing.

7.5   Wound healing and implants

Changes in inflammation and wound healing are important in implantation procedures since every intervention causes them to appear. It is always necessary to distinguish the changes due to the material from the underlying healing process. In any incised wound it has been shown that the changes due to surgery will last for at least 2 weeks; only after that can it be assumed that the material is responsible. The following factors associated with implanted material will significantly alter the normal pattern of wound healing.

The movement of fluid and chemical agents will be different depending on whether the implant is a solid or whether it is porous.

1. The early stage, with polymorphs, will be longer if there are toxic leachables, for example, plasticisers in some polymers.
2. If the implant moves in the tissue and causes damage, the cellular phase will be extended for as long as it continues. In addition to macrophage activity, it is when this happens that giant cells, which are not seen in normal wound healing, can be produced.
3. New blood vessels, critical in the healing process, can only approach a solid surface but may invade a porous one.
4. If the implant degrades, intentionally or not, chemically or mechanically, the cellular phase will be extended and giant cells appear.
5. If the surface is, or becomes, rough, macrophages and giant cells will remain until it becomes smooth.
6. Until the macrophages leave the site, the fibrous phase will be delayed.
7. During the final remodelling stage, the collagen is laid down along the lines of stress, which are controlled by the shape, size and properties of the implant. It is at this stage that the nature of the capsule around the implant is decided, its thickness and shape and whether adherent or not.

7.6   Implant–tissue interactions

It is clear that no implanted material is ever completely inert, its presence invariably affects the process of normal wound healing. The tissue response allows materials to be divided into four main categories, depending on the response.

1. If the material is toxic and either the material or something derived from it kills cells and surrounding tissue, it is unacceptable in any implant application. Occasionally an unanticipated toxic effect may derive from a leachable in the material but knowledge of the chemistry and an understanding of the published literature should prevent implantation of toxic materials. However, in some tests required by regulatory authorities, a control material of known toxicity may be needed so that effects can be quantified.
2. If the material is nearly inert, a fibrous, non-adherent capsule will form around it: the thinner the capsule, the more successful the material. Silicone rubber is a nearly inert material.
3. If the material is bioactive, an adherent interfacial bond occurs and there is minimal, if any, encapsulation. Bioactive glasses are examples of such materials.
4. If a material is biodegradable, dissolution occurs and the material is replaced by host tissue. It is essential that the products of dissolution be non-toxic and easily metabolised and that the structural integrity of the site is compatible with the dissolution rate. The materials used for degradable sutures are examples of such materials.

Many factors affect interfacial tissue response and all are mediated by the cells. Any material, which in its intended application produces minimal changes in the tissue, can be described as biocompatible.

Every application of a biomaterial enforces different conditions and a material may or may not be biocompatible in different applications. An enormous variety of biomaterials for implantation are now available and an understanding of the tissue response enables the biomaterials scientist to select the one that will function optimally in a given application. *A biocompatible material is one that possesses the ability to perform with an appropriate host response in a specific application.* [D.F. Williams, *Dictionary of Biomaterials*, p. 40, Liverpool University Press, 1999.]

7.7   Summary

Inflammation during healing of a wound is a natural consequence of implantation of any biomaterial. There are three stages of normal wound healing: the cellular phase, the fibrous phase and the maturation phase. Disruption of any of these phases, by release of toxic chemicals or micromotion, will lead to chronic inflammation and the presence of giant cells, in a wound that never heals completely.

7.8   Reading list

Alberts B., Bray D., Lewis J., Raff M., Roberts K. and Watson J.D., *Molecular Biology of the Cell*, 3rd edition, New York, Garland Publishing, Inc., 1994.

Bloom W. and Fawcett D.W., *A Textbook of Histology*, 10th edition, W.B. Saunders Co., Philadelphia, 1975.

Karp G., *Cell and Molecular Biology: Concepts and Experiments*, New York, J. Wiley & Sons, Inc., 1996.

Pollack G.H., *Cells, Gels and the Engines of Life*, Seattle, L. Ebmer & Sons, 2001.

Ratner B.D., Hoffman A.S., Schoen F.J. and Lemons J.E., *Biomaterials Science: An Introduction to Materials in Medicine,* 2nd edition, New York, Academic Press, 2004.

Part II

Clinical needs and concepts of repair

8

The skeletal system

LARRY L. HENCH

Imperial College London, UK

8.1   Introduction

Some 208 skeletal bones act as the framework and 501 separate voluntary muscles provide the co-ordinated power that allows humans to walk upright, run and take charge of their environment. The basic features of bone and other skeletal structures are described in this chapter. Bones generally act as a system of levers linked to each other by moveable connections, or joints. Joints vary widely in their structure and function. In the limbs the joints are highly mobile and lined with articular cartilage. The structure and biomechanical properties of articular cartilage are presented in a later section. The spinal column's movements are limited but sufficient, because of the way that the vertebral bodies are linked together. The bones of the cranial vault are jointed, but completely immobile.

The power and precision of the locomotor system is demonstrated by the act of picking something up. Using locating information transmitted via optic pathways, the motor cortex co-ordinates shoulder, elbow and wrist movements until the extended fingers touch.

Contact is signalled by touch receptors, and the joint positions then hold while the fingers grasp the object and secure it. An increase in muscle power is then required to lift it, but with sufficient relaxation to allow the joints to move. This muscular co-ordination depends on the integrity of the cerebellum, which is co-located with the base of the brain. It receives a constant stream of information from a rapid 'feedback' system. In addition, the cerebellum maintains muscular tone so that, even at rest, a muscle is ready for action. The final destination for its nerve impulses is the neuromuscular junction, a specialised structure on each muscle fibril. Acetylcholine released by the nerve endings crosses the gap to cause depolarisation of the membrane. A spreading wave of chemical change causes the actin–myosin molecular links to shorten, and the muscle contracts. Muscular activity greatly increases the simple weight-bearing stresses on bony tissues. Muscles are attached to the skeleton by means of tendons and ligaments, which transfer the muscular contractile motion to the skeleton.

In order to withstand stresses applied to the skeleton, bones have evolved to possess considerable strength and toughness. The tensile strength of bone is comparable to that of cast iron. A long bone like the femur (Fig. 8.1, CD Figs. 8.5, 8.11, 8.18, 8.19) shows cortical thickening to resist torsional strains in mid-shaft; at the hip the forces are mainly compressional, so the internal architecture is finely trabeculated (CD Figs. 8.11, 8.12, 8.19) along stress lines to reinforce and underpin the relatively thin cortex. The bones also constitute a mineral reserve for the body; disorders of calcium metabolism such as rickets and osteomalacia lead to characteristic deformities unless properly treated. The complexity of the bony structures within the locomotor system is well demonstrated by CD Fig. 8.18, which shows the complete skeleton of an adult. The process of ossification has followed that of growth. CD Fig. 8.19 shows the detailed structure of a long bone.

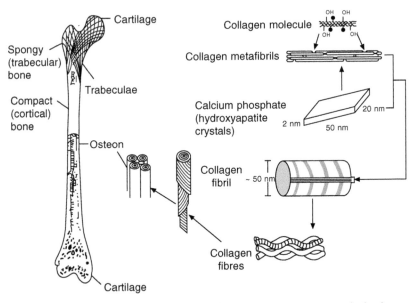

8.1 Schematic of the different bone structures or morphologies (CD Fig. 8.5).

8.2   The structural components of bone

The objectives of this section are to:

- describe the constituents of bone and their organisation into a composite microstructure; to compare the microstructure of different types of bone;
- relate the physical properties of density and strength of bone to microstructure;

- establish the effects of age, osteoporosis, strain rate and fatigue on the strength of bone and tendency to fracture.

Bone is composed of three major constituents:

1. living cells (osteoblasts, osteoclasts, and osteocytes);
2. non-living organic (collagen, muco-polysaccharides);
3. non-living inorganic crystals (hydroxycarbonate apatite, (HCA).

Osteoblasts are 'bone-growing cells'. Osteoclasts are 'bone-resorbing cells'. Osteocytes are mature bone cells surrounded by HCA bone mineral (CD Fig. 8.1).

Collagen is a very tough protein that makes up a large fraction of the structure of skeletal connective tissues (CD Fig. 8.2). There are 13 types of collagen in the body, as listed in Table 8.1. Type I collagen is the most important type of collagen in bone because it can be mineralised.

At the electron microscopic level collagen appears as a banded fibril due to its supramolecular organisation created from 300 nm collagen molecules. The four levels of structure of collagen fibrils are shown in CD Fig. 8.3. The

Table 8.1 Types of collagen

Type	Tissue	Polymeric form
Class 1		
(300 nm triple helix)		
Type I	Skin, bone, etc.	Banded fibril
Type II	Cartilage, disk	Banded fibril
Type III	Skin, blood vessels	Banded fibril
Type V	With Type I	Banded fibril
Type XI (1α, 2α, 3α)	With Type II	Banded fibril
Class 2		
(basement membranes)		
Type IV	Basal lamina	Three-dimensional network
Type VII	Epithelial basement membrane	Anchoring fibril
Type VIII	Endothelial basement membrane	Unknown
Class 3		
(short chain)		
Type VI	Widespread	Microfilaments, 110 nm banded aggregates cross-linked to Type II
Type IX	Cartilage (with Type II)	Unknown
Type X	Hypertrophic cartilage	Unknown
Type XII	Tendon	Unknown
Type XIII	Endothelial cells	

triple helix tertiary structure of collagen gives it high tensile strength and great flexibility. The Young's modulus (stiffness) of collagen is quite high (1–2 GPa) as is its tensile strength of 50 to 100 MPa, depending upon type and test method. CD Table 8.2 compares the Young's modulus and tensile strength to stiffness ratio with other skeletal tissues. Crystals of bone mineral form within and between type I collagen fibrils in a process called 'bone mineralisation' (CD Figs. 8.4, 8.5, 8.6). The crystals are a carbonated form of hydroxylapatite (HA) and are composed of CaO, P_2O_5 and OH molecules. The chemical formula of HA is $Ca_5(PO_4)_3OH$. The crystals are aligned along the axis of the collagen fibrils and reinforce the collagen matrix to provide a very strong and tough composite.

8.3   Microstructural features of bone

The constituents of bone are organised into three-dimensional structures. The several types of bone structures or morphologies, illustrated in Fig. 8.1, differ in their relative proportion and organisation of collagen and bone mineral. Woven bone, also called immature bone, is the weakest and cortical (also called compact) bone is strongest. Cancellous bone (also called trabecular or spongy bone) is intermediate in properties, as summarised in Table 8.2. The mechanical properties of articular cartilage and tendon are included in CD Table 8.3 for comparison. The mechanical properties of cartilage, tendons and ligaments differ greatly from bone because they are not mineralised and the proportion of cells, fibres, matrix and water are different, as summarised in CD Table 8.4 and CD Figs. 8.2, 8.7 and 8.8.

Because of a dense structure and the reinforcing effect of apatite crystals, cortical bone has a much higher modulus of elasticity than other skeletal tissues, Table 8.2 and CD Fig. 8.9. The orientation of the apatite crystals within and along the collagen fibrils creates structural units called 'osteons' (Fig. 8.1, CD Fig. 8.5). The osteons are oriented parallel to the long axis of

Table 8.2 The mechanical properties of skeletal tissues

Property	Cortical bone	Cancellous bone	Articular cartilage	Tendon
Compressive strength (MPa)	100–230	2–12		
Flexural, tensile strength	50–150	10–20	10–40	80–120
Strain to failure	1–3	5–7	15–50	10
Young's (tensile) modulus	7–30	0.5–0.05	0.001–0.01	1
Fracture toughness (K_{1c})	2–12			
Compressive stiffness (N)			20–60	
Compressive creep modulus			4–15	
Tensile stiffness (MPa)			50–225	

a bone (CD Fig. 8.10) and are responsible for anisotropic mechanical properties. Trabecular (spongy, cancellous) bone is much less dense than cortical bone, as shown in Table 8.2 and illustrated in CD Figs. 8.5, 8.11 and 8.12. Consequently, trabecular bone has a very much lower compressive strength than cortical bone, a lower tensile strength and a lower modulus of elasticity (Table 8.2 and CD Fig. 8.9). The anisotropic orientation of osteons is illustrated in a model in CD Fig. 8.13. The blood supply to the osteocytes via the Haversian system is also indicated. Optical micrographs of the osteons in compact cortical bone are shown in CD Fig. 8.14 with polarised light. Growth of osteons occurs as concentric rings of mineralised collagen around the osteocyte. The osteons intersect at a cement line. This process of growth is illustrated by various fluorescent labels given to at weekly intervals, as shown in CD Fig. 8.15. Structural details of bone as a living composite are illustrated in CD Figs. 8.16 and 8.17.

8.4   Biomechanics of bone: anisotropy of bone properties

The strength of cortical bone depends on whether it is loaded in tension, compression or torsion. The difference is due to the anisotropic, oriented structure of the osteons. Typical stress–strain curves for uniaxial tensile and compressive loading of human cortical bone are shown in Fig. 8.2.

The data show that cortical bone is stronger in compression than in tension. The strain to failure is about 3% for cortical bone under compressive loading

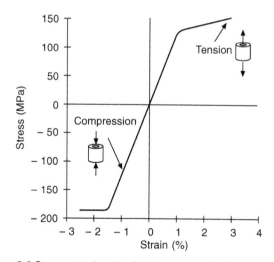

8.2 Stress–strain plot for human cortical bone for tensile and compressive loading. Data are shown for a longitudinal loading direction, CD Fig. 8.20.

regardless of whether the load is applied transversely across the bone or longitudinally along the bone. However, when the bone is loaded transversely in tension, the strain to failure and strength is reduced substantially. This anisotropic behaviour appears to be a structural response to the loading conditions that occur on the femur during everyday activities, such as walking. Specimens removed from a long bone and tested in tension illustrate the importance of structure on mechanical behaviour, i.e. transverse specimens are much weaker and exhibit much lower strain-to-failure than longitudinal specimens (CD Fig. 8.21).

8.5   Effect of age on bone

The effect of age on the bone mass of men and women and the changes in cross-section that occur to a long bone during ageing are substantial: as much as 60% for women and 50% for men, as shown in CD Figs. 8.22 and 8.23. The decrease in bone density greatly decreases the strength of trabecular bone, as illustrated in CD Fig. 8.24. The age-related effects on the strength at failure and elastic modulus of human femoral bone are less than on trabecular bone (CD Fig. 8.25) but are still important because the thickness of the cortical region decreases with age, as shown in CD Fig. 8.23. Microradiographs of trabecular bone of a normal 30-year-old female compared with trabecular bone of an osteoporotic 60-year-old woman (CD Fig. 8.42) show that nearly half of the trabeculae are lost and all are much thinner in the osteoporotic bone. A consequence of deterioration of trabecular bone density in the neck of a femur is fracture of the femoral head (CD Fig. 8.43). A consequence of osteoporotic thinning of cortical bone is the fracture of long bones (CD Fig. 8.44). Osteoporosis can have a severe effect on the trabecular bone density of vertebral bodies (CD Fig. 8.45). A clinical consequence is vertebral collapse (CD Fig. 8.46).

8.6   Effect of strain rate on bone

Bone is a viscoelastic material due to the properties of collagen and its composite structure. Consequently, both the elastic modulus and the strength of bone vary considerably with strain rate of loading. Bone is weaker under very slow rates of loading, as shown in CD Figs. 8.26 and 8.27. Rapid shock loading triples the stress-to-failure and reduces the strain-to-failure by about a factor of two. The elastic modulus increases with increasing strain rate (CD Table 8.5 and Fig. 8.27). Comparison of strain rate sensitivities for modulus and ultimate tensile strength of human cortical bone for longitudinal loading shows that, over the full range of strain rates, strength increases by about a factor of three, and modulus by a factor of two.

8.7   Fatigue failure of bone

CD Figs. 8.30–8.33 summarise the effects of fatigue on the failure of bone. Elastic modulus degrades with fatigue loading (CD Fig. 8.31). The number of cycles that bone can withstand before failure is shown in CD Fig. 8.32. Vigorous activity can reduce fatigue by a factor of 100. As with creep behaviour, resistance to fatigue fracture is greater for compressive loading. The number of loading cycles (n) required to fail the specimen is plotted against the maximum stress level (σ) attained during the cyclic test. This is known as a σ-n curve. The endurance limit is the stress below which cyclic fatigue of the material will not occur. Details of the fatigue behaviour of bone are given in CD Figs. 8.30–8.33.

8.8   Fracture of bone

CD Fig. 8.34 summarises typical modes of fracture for longitudinal bones; there are nine main fracture groups. The energy required to fracture bone is quite high, as shown in CD Table 8.6. As expected from age-related structural changes in bone, age decreases the energy required for fracture by two to three times.

8.9   Structure of tendons and ligaments

Tendons and ligaments are soft connective tissues that connect bones to bone, bone to muscle or support viscera. The structure of tendons and ligaments is that of a composite material composed of collagen fibrils embedded in an extracellular matrix composed of proteoglycans and elastin. Fibroblasts are the dominant cell type and are arranged in parallel rows between aligned bundles of primarily Type II collagen, as shown in Fig. 8.3 (CD Fig. 8.47).

The structure of collagen fibrils is shown in CD Fig. 8.3. The relative proportion of Type II collagen in tendon is 86% (dry weight), whereas for

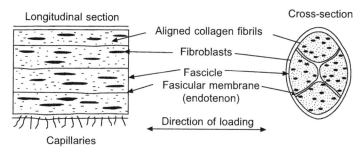

8.3 Schematic diagram of the structure of tendons and ligaments (CD Fig. 8.47).

ligament it is less, 70% (dry weight). The longitudinal bundles of collagen fibrils in tendon are bound together with glycoproteins and water to form fascicles. A loose connective tissue membrane, called the fascicular membrane or endotenon, separates the fascicles. The membrane permits longitudinal motion and supports blood vessels, lymphatic and nerves.

Ligaments contain less collagen, which is more randomly organised than in tendons. The collagen fibres in a tendon are aligned along the long axis of the tendon, whereas fibres in a ligament have a more woven pattern. Throughout the ligament structure is a uniform microvascularity. Although there is limited blood flow in a ligament, it is sufficient to provide nutrition to the fibroblasts and maintain matrix synthesis and repair. Damage to the microvasculature by tearing of the ligament insertion into bone, for example, will often lead to rupture, a common sports injury.

The blood supply of some tendons comes from many small capillaries that insert into a paratenon or mestotenon layer, referred to as an avascular tendon; nutrition of the fibroblasts comes from a synovial fluid, diffusional pathway. Repair of damaged avascular tendons is difficult because of the limited pathway for macrophages and cell proliferation.

8.10   Mechanical behaviour of tendons and ligaments

Tendons are anisotropic composite materials; ligaments have a somewhat lesser degree of anisotropy. The longitudinal alignment of collagen fibres along the direction of tensile forces gives rise to a very high tensile strength for tendons. The stress–strain load–deformation relationship of tendons, ligaments and skin are similar, as illustrated in Fig. 8.4 (CD Fig. 8.48).

At low levels of stress, tendons and ligaments stretch (strain) easily. This is called the 'toe' portion of the stress–strain curve and is due to the straightening of crimped collagen fibrils and orientation of fibres along the direction of applied load. With higher levels of stress, the highly oriented collagen fibres respond with a linear level of strain. The slope of the linear region represents the elastic modulus of the tendon or ligament. Values of elastic moduli for human tendons range from 1.2 to 1.8 GPa.

Deformation of tendons or ligaments loaded in their linear region is usually reversible, although the loading cycle has a hysterisis behaviour, as shown in Fig. 8.5. The area between the load–unload curves represents the energy loss in the tissue. This loss is low, only about 4 to 10% per cycle. Repetitive cyclic loading and unloading gradually moves the stress–strain to the progressively greater amounts of strain for less loads, a behaviour called creep. The decrease in stiffness of tendons allows it to lengthen gradually under cyclic loading and decreases the rate of muscle fatigue for a tendon–muscle unit. After about ten cycles the stress–strain response of a tendon

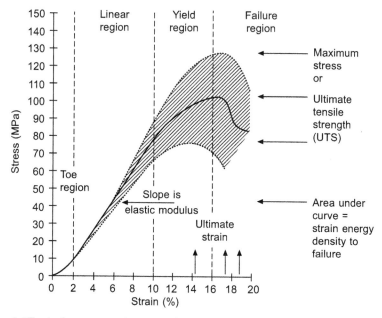

8.4 Typical stress–strain curves for tendons and ligaments (CD Fig. 8.48).

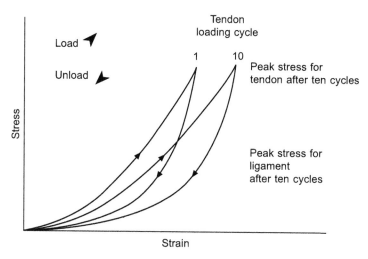

8.5 Effect of cyclic loading on stress–strain behaviour of a tendon.

usually becomes repeatable and steady. In contrast, cyclic loading of ligaments leads to a progressive decrease in the peak stress. After a large number of cycles, the ligament structure and mechanical response stabilise.

Continuing increases in level of stress applied to tendons and ligaments leads to irreversible changes at the interface between collagen fibres in the structure. Permanent stretching occurs, as represented by the yield region in the stress–strain curve (Fig. 8.4 (CD Fig. 8.48)). Thus, a tendon or ligament can stretch two to three times as much for a given level of load in the yield region as in the linear region.

Breakage of collagen–collagen or collagen–elastin bonds leads to fracture and failure of the tendon or ligament, the failure region of Fig. 8.4. The maximum stress or ultimate tensile strength (UTS) ranges from 45 to 125 MPa for human tendons. The ultimate strain to failure for human tendons is in the range of 9% to 35%. Ligaments have similar mechanical properties.

The large variation of biomechanical properties of tendons and ligaments is due to numerous factors, including anatomic location, the amount of activity (exercise), strain rate applied, and especially age.

8.11   Cartilage

Cartilage is also a soft connective tissue with important structural functions. There are three types of cartilage in the human body: elastic cartilage (ear and nose), fibrocartilage (intervertebrael space) and articular cartilage (at the articulating ends of bones which comprise synovial joints). Articular cartilage is especially important because it protects the underlying bone in a joint from abrasion and wear. The lamellar composite structure of articular cartilage, together with its synovial fluid lubricant, has a low coefficient of friction. Damage to articular cartilage, such as from trauma or osteoarthritis, exposes the underlying bone to stress concentrations and leads to pain from the nerves in the bone. Cartilage does not contain nerves. Self-repair of cartilage is difficult because of the low cellular content and absence of a blood supply. Nutrition of the cells comes from nutrients in the synovial fluid, which diffuse through interstitial water in the cartilage.

The structural features of articular cartilage include cells, called *chondrocytes*, encased in an extensive extracellular matrix composed of:

- collagen fibres (primarily Type II collagen with minor amounts of Type VI, IX, X, XI, XIV collagen, see Table 8.1);
- proteoglycans (chondroitin sulphate, keratan sulphate and hyaluronan);
- large amounts (65–80%) of interstitial water associated with the collagenous network and proteoglycan domains.

Chondrocytes differ in shape and function depending on their depth from the articular surface (CD Fig. 8.55). Near the surface the cells are flattened and aligned parallel to the surface (the superficial zone). In the transitional zone the cells are elliptical in shape and orientated obliquely to the joint surface. The third layer, called the deep zone or the proliferative zone, contains groups of four to eight spherical cells, capable of cell division, arranged in columns perpendicular to the surface. Underneath lies a calcified zone that is a transition to calcified cartilage and the underlying bone. Most of the cellular synthetic activity required to maintain the structure and function of articular cartilage comes from the transition and deep zones.

The unique viscoelastic mechanical properties of cartilage are due to a physicochemical interaction of the extracellular components. There is an equilibrium between the osmotic swelling pressures of the water-filled high molecular weight proteglycan gel which is balanced by the hydrostatic pressures due to tensile stresses in the cross-linked collagen fibre network. When cartilage is loaded in compression, the applied hydrostatic pressure creates a net pressure differential, which causes interstitial fluid to migrate away from the loaded area.

The forces transmitted through load-bearing joints are largely compressive. Articular cartilage has low frictional properties and thus is loaded mainly in compression perpendicular to the articulating surface. Collagen fibres in cartilage experience a tensile force under these conditions. The collagen fibres are aligned parallel to the surface in the superficial zone and perpendicular to the surface of the joint in the deeper zone (CD Fig. 8.55). The tensile stress–strain relationships of the collagen fibres reflect this difference in orientation, as illustrated in CD Fig. 8.55. Changes in orientation of the fibres with stress are resisted by the cross-linked proteoglycan gel and interstitial water and give rise to the viscoelastic behaviour of cartilage.

Age and osteoarthritis lead to loss of articular cartilage from bone surfaces. Small fissures appear in the superficial zone, exposing and disrupting the orientated collagen fibres, called fibrillation. The water content and stiffness decreases, microfracture occurs leading eventually to exposed bone, pain and deterioration of the lubricating characterisation of the joint.

8.12   Summary

The unique mechanical properties of bone, tendons, ligaments and cartilage reflect the highly anisotropic orientation of the cells, extracellular matrix and water in these connective tissues, as does the supply of nutrients. Age, trauma and disease degrade the structural constituents and nutritional supply of skeletal tissues and often lead to the need for implants to restore function, as described in Chapters 12 and 13.

8.13   Reading list

Bloom W. and Fawcett D.W., *A Textbook of Histology*, 10th edition, Philadelphia, W B Saunders, 1975.

Hughes S.P.F. and McCarthy I.D., *Sciences Basic to Orthopaedics*, Philadelphia, W B Saunders, 1998.

Revell P., *Pathology of Bone*, Berlin, Springer Verlag, 1986.

Simon S.R., *Orthopaedic Basic Science*, Illinois, American Academy of Orthopaedic Surgeons, 1994.

9

The cardiovascular system

M. J O H N   L E V E R
Imperial College London, UK

9.1   Introduction

The blood circulation in the body is essential for the delivery to the tissues of all the nutrients that they require, including oxygen, the removal of waste products, the control of fluid levels, the control of multiple functions through the endocrine hormone system and for thermoregulation. The three main components of the cardiovascular system are the blood, the blood vessels and the heart. Blood is the fluid that is in continuous movement through all tissues of the body, blood vessels are the conduits for the blood and the heart is the muscular pump that is primarily responsible for driving the flow.

Blood is a suspension of cells in a fluid called the plasma. The cells are mainly erythrocytes (red blood cells), which have a gas transport function. The plasma contains proteins, ions, nutrients, metabolites, waste products and signalling chemicals. It is the movement of blood around the body that ensures the efficient transport of materials and heat to ensure the maintenance of all tissues. The other main function of the blood is a protective one, which comes into effect when the body is injured or is invaded by pathogens (see Chapter 7). Apart from the erythrocytes, the other cellular components are leukocytes and platelets. Leukocytes are involved in inflammatory and immunological processes, and platelets have a role in haemostasis when blood vessels are damaged.

Blood vessel types

Blood vessels change progressively in size and properties throughout the system. The arteries leading from the heart are large, tough and compliant but become progressively smaller and stiffer as they branch away from the centre. Their main components are elastin, which is a protein that confers compliance on the vessels, and smooth muscle, which is also compliant but can also contract. Arteries lead to arterioles, whose contractile properties are important in controlling blood flow to the organs that they supply and providing a resistance against which the beating heart can maintain an adequate pressure to ensure flow to all parts of the body. Capillaries are the smallest blood vessels, commonly being only about 5 μm in diameter, which is smaller than the blood cells that have to squeeze through them. Capillary walls are normally only one cell in thickness, allowing optimal exchange between the blood and the tissues through which they permeate. The density of capillaries varies between different tissues but is always very high, ensuring that materials only have to move a few tens of micrometres at most. From the capillaries, the blood drains into venules and veins of increasing size and is eventually returned to the heart. The veins have thinner walls than the arteries as the blood within them is normally at a much lower pressure. Many veins have semi-lunar, one-way valves that prevent backflow. The blood flow decelerates dramatically in the arteries and microvessels and then accelerates as it converges into the venous system. The resulting variations in blood velocity ensure rapid transport around the body but also leave adequate time for the blood to spend in the capillaries to ensure near equilibration of gases, metabolites and waste products with the surrounding tissues. In all but the fetus, the pulmonary circulation system operates in parallel to the systemic cardiovascular system, ensuring that all blood is subject to gas exchange in the lung, while the systemic circulation supplies blood to all the other tissues in the body.

The heart

The heart is a four-chambered structure composed primarily of specialised muscle fibres that contract rhythmically. The contraction is initiated by depolarisation of cells in the pacemaker region followed by transmission of the action potentials via conducting fibres, first to the atria and subsequently to both ventricles. Contraction of the chambers raises the pressure of blood within them, forcing it through one-way valves into the aorta, the major artery. These valves, together with those in the veins, are responsible for the unidirectional flow of blood through the circuit. The heart is supplied with its own coronary circulation, which originates from the base of the aorta. More oxygen is extracted from coronary artery blood than from the circulation of most other organs, so any pathological change within it has very deleterious consequences.

9.2   Cardiovascular pathology

Failure of any of the components of the system can have catastrophic effects. Loss of normal blood flow to the brain for less than 1 minute causes loss of consciousness and, if maintained for a few minutes, causes irreversible damage and death. Other organs are more tolerant of temporary cessation of flow (ischaemia), but in all cases there will be progressive loss of function followed by necrosis (cell death). Even relatively short interruptions in flow can result in adverse reactions when flow resumes (reperfusion injury).

Although the rate has been declining in recent years, diseases of the cardiovascular system still account for well over 40% of all deaths in the developed world (Fig. 9.1, CD Fig. 9.2).

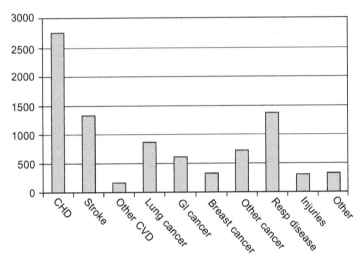

9.1 Death rates per million of population in England and Wales 2001. (Adapted from Office of National Statistics data 2002.) CHD = coronary heart disease, CVD = cardiovascular disease (CD Fig. 9.2).

Sadly though, the death rate from cardiovascular diseases has been rising dramatically in the developing nations mainly through the adoption of 'western' lifestyles. Figure 9.2 (CD Fig. 9.3) shows the transition that is occurring at the present time.

The commonest manifestations of cardiovascular disease are shown in Fig. 9.3. Coronary heart disease is the most prevalent cardiovascular condition, causing more deaths and long-standing disability than any other. The underlying cause is atherosclerosis. This condition develops very early in childhood, then normally regresses before a resurgence on reaching adulthood. It involves the deposition of material including fat, cell debris and connective tissues in the inner wall of blood vessels, particularly of the medium-sized arteries. The susceptibility of an individual is dependent on various 'risk factors', including family history, diet, blood pressure, smoking habits and exercise levels. On progression, there can be one of two serious consequences. Either the atherosclerotic plaque can increase in size, blocking the blood vessel and impeding blood flow past it (a stenosis) or the plaque may be ruptured by the haemodynamic forces imposed on the vessel wall. The contents of the ruptured stenosis may then form emboli that can block vessels downstream, which is a common cause of stroke. Alternatively, blood clots may form on the damaged tissue and either block the vessel locally or downstream. Coronary heart disease is the ischaemic condition in the heart muscle resulting from any of these processes. The consequence may be pain, angina, or at worst failure of the pump (myocardial infarction).

Because atherosclerosis is so prevalent, many of the different conditions in Fig. 9.3 may be present simultaneously; it is very rare for peripheral vascular disease to be present without the coronary arteries being affected.

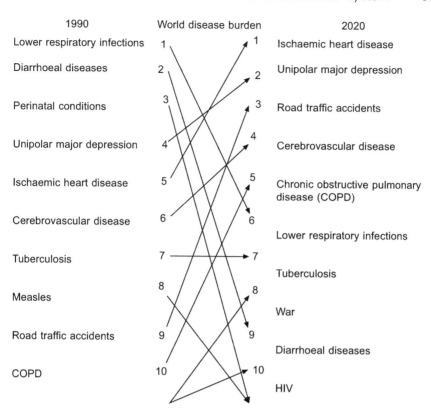

9.2 Changes occurring in the incidence of disease throughout the whole world. (Adapted from Lopez and Murray (1998), *Nature Med*, 4: 1241.) (CD Fig. 9.3).

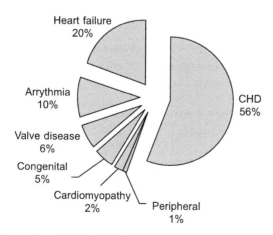

9.3 Incidence of cardiovascular diseases in the UK. Data from the British Heart Foundation. CHD = coronary heart disease (CD Fig. 9.4).

9.3   Control and treatment of cardiovascular pathologies

In the case of some cardiovascular pathologies, it has been possible to use pharmacological or surgical therapies to produce a perfectly functional system. For example, about 50 years ago, with the advent of open-chest surgery it began to be possible to correct congenital malformations of the heart, starting with reduction of patent foramina between chambers which had not closed at birth. More recently, more complex malformations including transposed great vessels or absent heart chambers can be treated this way. Similarly, it has been possible over the past 20 years to use drugs to control hypertension, which previously would have led to more serious complications such as heart failure. Increasingly though, implantation techniques are being employed to replace damaged or diseased components or to assist the system.

Blood replacement

Up until recently, transfusion of homologous, immunologically matched blood from blood banks was the normal procedure for dealing with severe blood loss. The quest for new materials for blood replacement initially arose because of inaccessibility of donor blood in wartime but has more recently been heightened by concerns about diseases such as hepatitis and acquired immune deficiency syndrome (AIDS). The feature of blood which is most important to match in artificial substitutes is its oxygen-carrying capacity. Consequently, the most successful approaches so far have been to use fluid suspension such as perfluorocarbons (in which oxygen has an unusually high solubility) or solutions or suspensions of haemoglobin derivatives.

Implants in the treatment of cardiovascular pathology

Devices such as pacemakers can aid the circulation without contact with the blood within it. Most implants, though, must not only satisfy the conditions of biocompatibility that apply elsewhere in the body but must also minimise clotting abnormalities.

Blood clotting is a very finely balanced process, able to cause very rapid haemostasis when vessels are damaged, but tuned so as not to allow spontaneous clot development within the vessels and thereby impair blood flow. There are two overlapping clotting cascade mechanisms: an extrinsic pathway dependent on tissue factors released from damaged cells and an intrinsic pathway usually initiated by the contact of blood platelets with abnormal surfaces. The choice of implanted materials is consequently very strongly dependent on their non-thrombogenicity as well as on other factors such as their chemical inertness.

9.4   Summary

Blood, blood vessels and the heart comprise the cardiovascular system (Figs. 9.4 and 9.5, CD Figs. 9.1, 9.5), which is essential to maintain life as the movement of blood cells and plasma throughout the body provides nutrition and removes waste products from cells.

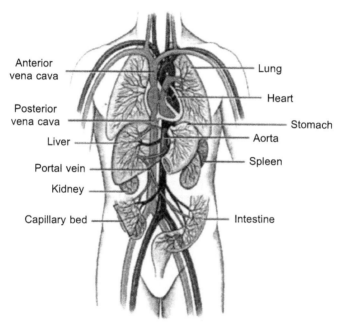

9.4 Schematic of the cardiovascular system. Modified from www.agen.ufl.edu (CD Fig. 9.1).

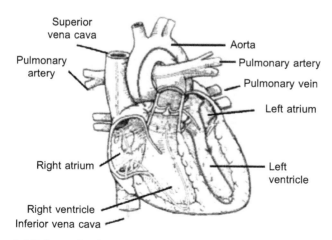

9.5 Schematic showing the structure of the heart and blood flow directions. (CD Fig. 9.5).

Failure of any components of the system can lead to irreversible damage to cells and organs, and is one of the leading causes of death in the developed world. Thus, repair and replacement of the components of the cardiovascular system is one of the great challenges of biomaterials and bioengineering.

9.5   Reading list

Borozino J.D. (ed), *The Biomedical Engineering Handbook* (Chapters 1, 21, 31, 33, 34, 46), Boca Raton, Florida, CRC Press, 1995.

Kauffmann-Zeh A. and Dhand R., Nature insight: vascular biology, *Nature,* **407**, 219–269, 2000.

Levick J.R., *An Introduction to Cardiovascular Physiology*, 4th edition, London, Arnold, 2003.

Lopez A.D. and Murray C.C.L.J., The global burden of disease, 1990–2020. *Nature Medicine,* **4**, 1241, 1998.

10

Biomedical polymers

ROBERT G. HILL
Imperial College London, UK

10.1 Introduction

Polymers that are biocompatible, i.e. those that are not toxic to the body on implantation, can be classified as being bioinert or bioresorbable. Generally, high molecular weight biocompatible polymers are non-degradable and are classed as bioinert. Toxicity can occur with normally biocompatible polymers due to leaching of low molecular weight plasticisers and additives. It is important to characterise the grade of polymer in use. What is sold as polymer X by one manufacturer may be very different from polymer X sold by another, due to purity and additives present. Surface reactions and absorption of proteins at the polymer surface can also cause problems. Therefore, the surface texture and the shape of the implant are also important.

10.2 Bioinert polymers

Common non-degradable medical polymers include: polyethylene terephthalate (PET), nylon 6,6 polyurethane (PU), polytetrafluoroethylene (PTFE), polyethylene (PE, low density and high density and ultra-high molecular weight, UHMW), polysiloxanes (silicones) and poly(methylmethacrylate) (PMMA).

It should be noted that there is some evidence of enzymatic degradation of PET, nylon and PU but the amount of degradation is generally very small, the exception being some types of polyurethanes.

Poly(methylmethacrylate)

Poly(methylmethacrylate) is a hard rigid, glassy but brittle, polymer with a glass transition temperature of about 100 °C. The chemical structure is shown in Fig. 10.1 (CD Fig. 10.1). It is classified as bioinert. In set forms it is used as intraocular lenses and hard contact lenses. *In situ* setting forms (known as cold curing) are used as bone cements in joint replacement surgery. CD Fig.

$$[-CH_2-\underset{\underset{\displaystyle CH_3}{\overset{\displaystyle |}{O}}}{\overset{\underset{\displaystyle |}{\overset{\displaystyle |}{C=O}}}{\overset{\displaystyle |}{\underset{\displaystyle CH_3}{C}}}}-]_n$$

10.1 Chemical structure of poly(methylmethacrylate) (CD Fig. 10.1).

10.2 shows a schematic of a total hip replacement. A viscous paste is mixed by the surgeon and used as a packing material between the metal femoral prosthesis and the internal bone (intermedullary) cavity. The paste hardens as methylmethacrylate (monomer) converts to poly(methylmethacrylate). During this conversion a 21% volume shrinkage occurs. To reduce the shrinkage, pre-polymerised poly(methylmethacrylate) powder is mixed with monomer; the shrinkage is reduced to about 3%. This shrinkage is offset after polymerisation by water absorption and comcomitant expansion. However, this material is not ideal as a bone cement. The addition polymerisation is highly exothermic and temperatures in excess of 90 °C are reached, which can lead to thermal bone necrosis during total hip replacement.

Poly(tetrafluoroethylene) (PTFE)

PTFE has the chemical structure $[-CF_2-CF_2]_n$. It is chemically extremely stable and is a classic example of a bioinert polymer. It must be noted that all commercial PTFEs only approximate to the chemical composition given above. PTFE is highly crystalline and the crystallites have a high melting point (330 °C), which makes PTFE difficult to process. It cannot be moulded to shape. Particles are sintered then machined to the required form. The commercial material Gortex™ is a fibrous sheet form of PTFE that has numerous uses as a membrane material. PTFE has relatively poor mechanical properties with a low yield strength, which limits its use to non-structural applications. It is used as part of vascular prostheses in the form of coatings on Dacron®.

Polyethylene

Polyethylene has the chemical structure $[-CH_2-CH_2]_n$. Three types are used in biomedical applications:

- low density polyethylene LDPE (lower degree of crystallinity);
- high density polyethylene HDPE (higher degree of crystallinity);
- ultra-high molecular weight polyethylene UHMWPE (molar mass $> 10^6$).

For most structural applications PE has too low a yield strength. The yield strength increases with molecular weight and degree of crystallinity. UHMWPE is used as the bearing surface, the acetabular component of total hip replacements, and is only used in orthopaedics. LDPE and HDPE are readily mouldable. UHMWPE is not, and, like PTFE, is sintered and machined to shape. Polyethylene, like PTFE, is a hydrophobic (water-repellent) and bioinert polymer. However, whilst polyethylene is classified as bioinert, UHMWPE particles in the submicrometre size range arising from wear of acetabular cups are very toxic and cause bone necrosis and osteolytic lesions, which are a major contributing factor to the aseptic failure of hip joints (Chapter 13).

Polysiloxanes (silicones)

Polysiloxanes are widely used for medical applications and have a long success record. Material types include elastomers, gels, lubricants, foams and adhesives. Polysiloxanes are very chemically stable and unreactive. They are very hydrophobic and have a low moisture uptake. They have good electrical insulation characteristics. Polysiloxanes are the polymer of choice for long-term use in the body where an elastomer is required and where there is a demand for biodurability and biocompatibility. Nearly all polysiloxanes are based on polymethylsiloxane; the chemical structure of which is given below:

$$\begin{array}{c} CH_3 \\ | \\ [-O-Si-]_n \\ | \\ CH_3 \end{array}$$

Note the lack of any polar groups in the structure. This leads to a very hydrophobic polymer with poor wetting characteristics. Polymethylsiloxane is rarely used without modification.

Polysiloxane elastomers

These consist of a cross-linkable modified polysiloxane of high molar mass ($>3.5 \times 10^5$), a reinforcing filler and a catalyst system to initiate cross-linking. These materials only flow easily under pressure and are fabricated by techniques such as compression moulding, transfer moulding, calendering and extrusion. Low molecular weight grades($< 10^5$) can be fabricated by solvent casting and reactive injection moulding. Vulcanisation/cross-linking systems may involve benzoyl peroxide, platinum catalysts and room temperature moisture curing systems (RTVs).

Polysiloxane gels

These are similar to the elastomers, but contain no filler and are generally based on a low molar mass polymethylsiloxane that is very lightly cross-linked. These are used in breast implants. Some problems have occurred with uncross-linked low molar mass chains and impurities leading to 'bleeding', where the low molar mass siloxanes permeate the envelope by osmotic driven diffusion. They are not recommended for use except when contained by an impervious envelope.

Polysiloxane adhesives

There are two basic types of polysiloxane adhesive:

- There are those that cure and crosslink on contact with water, e.g. Silastic®️ Medical Adhesive type A. This type contains an acetoxy ligand that reacts with water to form silicon–oxygen–silicon cross-links and acetic acid. There is some evidence of a slight reaction to the acid produced when leached.
- There are those that rely on their 'stickiness' only for adhesion, e.g. Dow Corning®️ 355. This type is contained within a fluorocarbon solvent. It can be used to affix materials to the skin, such as metals, paper, glass, fabric, polysiloxane and elastomers. Its adhesive properties to polysiloxanes are not good over long time periods because the solvent diffuses into the polysiloxane.

Polyamides (Nylon®️)

Nylons are based on polymers containing a –CONH– linkage. The major types used in medicine are based on Nylon 6, Nylon 6,6 and Nylon 6,12, where the numbers indicate the number of carbons separating the amide linkages. Both fibres and mouldings are partially crystalline and the degree of crystallinity influences the properties. Nylons are fairly hydrophylic and the amide linkage can form hydrogen bonds with water molecules in the amorphous regions, resulting in significant water uptake. Water uptake results in plasticisation and a marked change in mechanical properties.

Poly(ethyleneterephthalate) PET

Poly(ethyleneterephthalate) has the structure given in Fig. 10.2 (CD Fig. 10.3). PET is a relatively hydrophylic polymer but is more hydrophobic than nylons. Protein absorption and thrombogenicity (creation of blood clots) increases with hydrophobic character. PET is better for blood contacting devices than polyamides for this reason and because polyamides have a

$$\left[-CH_2-CH_2-\underset{\displaystyle}{\bigcirc\!\!\!\!\bigcirc}-\overset{\displaystyle O}{\underset{\displaystyle}{\overset{\displaystyle \|}{C}}}-O- \right]_n$$

10.2 The chemical structure of poly(ethyleneterephthalate) (PET) (CD Fig. 10.3).

tendency to degrade. The commercial material Dacron® is based on amorphous PET fibres. PET for blood contacting devices is often used as woven or knitted tubes. The tightly knit forms support maximal cell coverage whereas loose knits do not support tissue in growth.

Polyurethanes

Polyurethanes are polymers that contain the urethane group.

$$-\overset{\displaystyle H}{\underset{}{\overset{|}{N}}}-\overset{}{\underset{\displaystyle O}{\overset{}{\underset{\displaystyle \|}{C}}}}-O-$$

A large number of urethane polymers exist with widely different physical and biological properties. The urethane grouping can be considered as resulting from the reaction of an isocyanate and an alcohol:

$$R'-N=C=O + R-OH \rightarrow R'-\underset{\displaystyle H}{\overset{}{N}}-\underset{\displaystyle O}{\overset{}{C}}-O-R$$

The polyurethane is then represented by:

$$[-R-O-\underset{\displaystyle O}{\overset{\displaystyle \|}{C}}-NH-R'-HN-\underset{\displaystyle O}{\overset{\displaystyle \|}{C}}-O-]_n$$

There is a wide variety of polyurethanes due to the numerous possibilities for R and R'. R is typically an oligomeric (molar mass 200–500) hydroxyl-terminated polyether or polyester. Generally, synthesis is a two-stage process involving the preparation of a low molar mass pre-polymer followed by chain extension and/or cross-linking. Common pre-polymers are based on 2,4-toluene diisocyanate (TDI) or 4,4'-diphenylmethane diisocyanate (MDI). Polyether urethanes are usually based on polytetramethylene oxide (PTMO), polypropylene oxide (PPO) and polyethylene oxide (PEO). Polyester urethanes are typically based on polycaprolactones.

Chain extension may be performed by glycols or diamines. The nature of the chain extender is very important in that it determines chain flexibility and modulus. It is also possible to have urea groups –NH–CO–NH– and the urethane group –NH–CO–O–. Most polyurethanes for medical use are two-phase block copolymers (also termed segmented polyurethanes). The polyester or polyether glycol forms the soft segment and matrix phase.

In the early use of polyurethanes, acute reactions were observed on implantation. Most of the studies were done on commercial products rather than on well-characterised materials. The reactions observed are now believed to be due to *in vivo* ester hydrolysis of polyester urethanes. Problems have also occurred with rejection of breast implants coated with polyester urethanes due to degradation of the urethane. Segmented polyether urethanes are preferred because of their greater stability and lack of susceptibility to hydrolysis. Commercial examples include Biomer®, Pellethane® and Tecoflex®. Segmented polyether urethanes have good haemocompatibility and are one of the preferred polymer types for blood contacting devices.

10.3   Bioresorbable polymers

When certain polymers are exposed to liquids such as body fluids, they may swell or dissolve. The small molecules of the liquid may diffuse into the polymer, pushing the chains apart and increasing the volume. This may occur preferentially at scratches on the surface, leading to a local tensile stress and result in crazing or environmental stress cracking. Dissolution may be viewed as an extreme case of the above.

Degradation by this route is reduced by cross-linking, by increasing molar mass, by increasing the degree of crystallinity and by reducing the temperature.

A bioresorbable polymer is designed to degrade within the body after performing its function. Useful materials often degrade to give normal metabolites of the body. Examples include: polylactide, polyglycolide, poly(-3-hydroxybutyrate), polyhyaluronic acid esters, polydioxanone and copolymers of the above, plus additional species such as poly(glycolic acid/lactic acid) and poly(glycolide-trimethylene carbonate).

Biodegradable/hydrolysable polymers are frequently the basis of scaffolds for tissue engineering. Tissue engineering is growth of tissue *in vitro*, often by seeding cells on a template (scaffold) that can guide the tissue growth (see Chapters 18–22). Other bioresorbable polymers include those explained below.

Polymers that are produced by a condensation route (see Chapter 4) are prone to hydrolysis, which is the reaction with water to produce –OH bonds. In addition, polymers may contain side groups that are capable of being hydrolysed. The rate of hydrolysis is dependent on the water absorption of the polymer and is often limited by the diffusion of water through the polymer. Diffusion of water in polymers is often related to their solubility parameter, their glass transition temperature and their degree of crystallinity.

Polyesters based on $(-R-COO-)_n$ are often susceptible to hydrolysis. Degradation is pH dependent. Esters are hydrolysed at a faster rate under acid and alkaline conditions than they are at neutral pH. Since on hydrolysis of an ester a carboxylic acid is produced, the pH falls during hydrolysis, which accelerates the degradation process.

Poly(α-hydroxy acid)s are a group of polymers whose repeating unit is based on $-(O-CO-CHR)-n$ and are derived from α-hydroxy acids HO–CHR–COOH. They have been under development for osteosynthesis (bone repair) devices since the 1960s.

Polyglycolide is widely used for degradable sutures. Implants have been fabricated from both polyglycolides and polylactides. Both polymers are polyesters and possess an ester group in the polymer backbone that can be hydrolysed causing chain scission. The degradation products of these two polymers are glycolic acid and lactic acid respectively, both of which occur naturally in the body.

Polyglycolide

Polyglycolide (PGA) is based on $(O-CO-CHR)-n$ where R = H. High molecular weight PGA is a hard, tough crystalline polymer melting at about 228 °C with a glass transition temperature of 37 °C. The polymer can be spun to produce fibres when the molar mass is between 2×10^4 and 1.45×10^5. The strength of PGA in the fibre direction is increased when it is spun into fibres, because of the preferred molecular orientation of the polymer chains. High molecular weight PGA is made by dehydrating hydroxyacetic acid and glycolic acid to form glycolide, which is the cyclic dimer condensation product. PGA can be synthesised from glycolide under the influence of metal salt catalysts at low concentration by a ring opening polymerisation. The molar mass of the polymer is determined by temperature, time, concentration of the catalyst and concentration of any chain transfer agents. The polymer is spun into fibres, which are then twisted together to form thread for sutures.

Polyglycolide or poly(glycolic acid) has a lower glass transition temperature and a higher water absorption than poly(-3-hydroxybutyrate) or PET. The pure polymer degrades very slowly. However, when it is co-polymerised with a small proportion of another polymer to prevent crystallinity, it degrades in about 20 days in the body and is used as degradable sutures (stitches).

Polylactides

Polylactides are based on $-(O-CO-CHR)-n$ where R = CH_3. Replacing H by CH_3 leads to a more hydrophobic polyester and results in lower water uptake and lower hydrolysis rates. Having R = CH_3 results in a chiral centre and leads to D and L forms as well as DL racemic forms where there is a random arrangement of chiral centres. D and L forms can crystallise whilst the racemic form cannot. The degree of crystallinity influences strength, fracture toughness and degradation behaviour. The crystalline regions do not take up much water and the crystalline forms are much more resistant to degradation. The crystalline forms can result in crystallites being left after degradation giving rise to particulates and a toxic reaction *in vivo*.

Degradation is often autocatalytic. Degradation of thick sections can occur faster than thin sections due to the build-up of a localised low pH accompanying degradation within the section and the formation of lactic acid. This can result in the rapid release of lactic acid and polylactide oligomers, resulting in a toxic response. This is overcome by using basic fillers that neutralise the acidic carboxyl groups produced on hydrolysis. The hydrolysis occurs randomly along the chain and the molecular weight reduces rapidly. If autocatalytic degradation does not occur, this leads to the polydispersity (M_w/M_n, see Chapter 4) going to two. The reduction in molecular weight results in a marked reduction in fracture toughness and strength.

Applications of polylactides include bone plugs, screws and fracture fixation plates. Applications are limited as a result of rapid reduction in strength *in vivo*.

Poly-3-hydroxybutyrate (PHB)

Poly(3-hydroxybutyrate) will be considered as an example of the potential of biotechnology for producing new materials and of the advantages of biological synthesis. It has many properties that are attractive for biomedical applications; in particular, it is a polyester-like polyglycolide and is also biodegradable within the body. Figure 10.3 (CD Fig. 10.3) shows the structure of PHB. Note the presence of a chiral centre is denoted by the asterisk. Since PHB is produced by bacteria, only one optically active form is produced and the configuration is absolutely perfect. This leads to PHB being obtainable with degrees of crystallinity of > 95%.

Like polyglycolide, PHB is degraded by hydrolysis but one of its major attributes is that it is similar to PET in terms of mechanical properties. It has a glass transition temperature of about 10 °C, but it is less hydrophobic than PET and so its water absorption is higher. Not only can it degrade in the

$$[-\underset{\underset{H}{|}}{\overset{\overset{CH_3}{|}}{C^*}}-CH_2-\overset{\overset{O}{\|}}{C}-O-]_n$$

10.3 The chemical structure of poly(-3-hydroxybutyrate) The * denotes the chiral centre (CD Fig. 10.3).

body, but it can also undergo degradation in soil, making it attractive as a degradable packaging material. Degradation in soil is facilitated by soil bacteria. The degradation product is hydroxybutyric acid, which like glycolic acid and lactic acid is a normal metabolite found in the body. However, PHB degrades too slowly for many biomedical applications, largely as a result of its high crystallinity, which hinders water diffusion into the polymer.

Hyaluronic acid

Hyaluronic acid is an important biopolymer that bridges the gap between synthetic and naturally occurring polymers, i.e. the gap between biochemistry and synthetic polymer chemistry. Hyaluronic acid is also an example of a commercially useful biopolymer with many existing and potential uses as a biomedical material. It is an important extracellular biological polysaccharide and plays a major role in regulating the environment in which many cells exist. Being a polysaccharide, it does not have a monodisperse molecular weight distribution and its structure is not determined by the genetic code. Furthermore, because it is a polysaccharide, it does not elicit an immune response.

Hyaluronic acid is a naturally occurring polysaccharide and therefore it is completely biodegradable *in vivo*, which makes it suitable as a matrix for controlled drug delivery and as a matrix for hybrid implants consisting of living cells in a hydrogel matrix. It is also a suitable material for non-adhesive wound dressings. It is important to note that most cells do not exist in crystalline matrices or free solution, but in hydrogels, and that the commonest hydrogel is based on hyaluronic acid (Chapter 11).

In addition to forming the largest part of the extracellular matrix, hyaluronic acid is also an integral part of the aqueous humour of the eye and the synovial fluid of the joints, where it exerts rheological control and serves as a viscoelastic damping medium. It controls the water content of the extracellular fluid, serving to resist bacterial penetration whilst allowing molecules to gain access to the cell and exit from the cell. It binds cations and strongly influences cell mobility and tissue regrowth. The properties of hyaluronic acid gels are determined by their molar mass, molar mass distribution and concentration, which largely determine the mesh density, permeability and mechanical properties.

The structure of hyaluronic acid is shown in Fig. 10.4 (CD Fig. 10.5). The properties of hyaluronic acid can be modified and, within the body, a large number of related polysacharides exist. Commercial hyaluronic acids are extracted from rooster combs, but hyaluronic acids can also be produced using biochemical engineering from genetically engineered bacteria. The molar mass of the hyaluronic acid can be varied to alter the viscosity and gel properties. Hyaluronic acids degrade in 2–5 days in the body without

10.4 Chemical structure of hyaluronic acid (CD Fig. 10.5).

modification. However, the basic hyaluronic structure can be modified by esterification with hydrophobic groups and by cross-linking.

Esterification increases the hydrophobicity, and both esterification and cross-linking reduce the water uptake, reduce the expansion and reduce the degradability. This enables the properties of commercial hyaluronic acids to be tailored to given applications.

10.4   Summary

This chapter introduces polymers that are biocompatible, i.e. they can be implanted into the body without a toxic response. Bioinert polymers such as polyethylene and polymethylmethacrylate are used in joint replacements (Chapter 13) and are designed to remain in place, unchanged for many years. Polyethylene has been used in composites such as HAPEX®, which has been used as middle ear prostheses (Chapter 5). Polyesters such as polyglycolic and polylactic acids are used as resorbable polymers in applications such as dissolving sutures. Resorbable polymers are also being investigated for use as scaffolds for tissue engineering applications (Chapter 19).

10.5   Reading list

Andrade J.D., *Surface and Interfacial Aspects of Biomedical Polymers*, New York, Plenum, 1985.

Gebelein C.G., *Advances in Biomedical Polymers*, New York, Plenum, 1987.

Shalaby W., *Bioabsorbable Polymers for Biomedical and Pharmaceutical Applications*, Lancaster, Pensylvania, Technomic Publishing Co, 2001.

Zaikov G.E., *Biomedical Applications of Polymers,* Hauppauge, New York, Nova Science Publishers Inc., 1999.

11

Biomedical hydrogels

JASON A. BURDICK
University of Pennsylvania, USA

MOLLY M. STEVENS
Imperial College London, UK

11.1 Introduction

Hydrogels are insoluble water-swollen networks that are being investigated for biomedical applications such as drug delivery and tissue engineering. These polymers consist of a wide range of chemistries (which will be detailed later in this section) and can be designed with a wide variety of material properties (e.g. mechanics and degradation). Table 11.1 (CD Table 11.1) presents a list of potential advantages and disadvantages to using hydrogels as biomaterials.

Table 11.1 Advantages and disadvantages of hydrogels as biomaterials

Advantages	Disadvantages
Wide range of chemistries available	Poor mechanical properties
Typically biocompatible	Difficult to sterilise
High water content	
Potentially injectable to be minimally invasive implantation	

11.2 Mechanisms of hydrogel formation

Hydrogels are formed from both small (monomers) and large (macromers) precursors through a variety of reactions. Additionally, hydrogels consist of homopolymers (one monomer), copolymers (more than one monomer), and semi-interpenetrating networks, where one monomer is polymerised throughout an already cross-linked network. By altering the type of hydrogel, various physical properties can be altered and molecules can be introduced that control the hydrogel's interactions with cells and tissues.

Hydrogels are formed through a variety of mechanisms, including physical and chemical gelation (illustrated in Fig. 11.1, CD Fig. 11.1 and 11.2). Physical gelation occurs when polymer chains are bonded through ionic

Polyanions

Cations or polyanions

Example: sodium alginate + calcium

Chemical gelation

Or

Initiator

radiation

Example: methacrylated poly(ethylene glycol)

Or

Multifunctional macromers

11.1 Schematic of various gelatine processes (CD Figs. 11.1 and 11.2).

interactions, hydrogen bonding, through molecular entanglements, or through the nature of the material's hydrophobicity. In general, these gels are heterogeneous due to the complexity of the polymer–polymer interactions. An example of physical gelation is shown in Fig. 11.1, where a polyelectrolyte (e.g. alginate, a polyanion) is mixed with either a molecule (e.g. Ca^{2+}) or a polymer of opposite charge (e.g. poly(L-lysine), a polycation) to form a hydrogel. Other examples include block copolymers of ABA or AB type, where A is a hydrophilic block and B is a hydrophobic block, or the formation of crystallisation within a polymer using repeated freeze–thaw cycles, e.g. poly(vinyl alcohol). Physical hydrogels can degrade either through the dissociation of the bonded or entangled chains or through degradable units found throughout the polymer backbone.

An alternative process involves chemical gelation, where hydrogel chains are covalently linked together. This process uses techniques such as radiation, the addition of chemical cross-linkers and the use of multifunctional reactive compounds. Again, the hydrogels formed are typically heterogeneous due to variations in the cross-linking density throughout the networks with pockets of highly cross-linked areas. As shown in Fig. 11.1, one example of chemical gelation is the polymerisation of methacrylated poly(ethylene glycol) in the presence of an initiator (either photo or thermal) and the appropriate light source or an increase in temperature. The reactive groups can be located at the end, as pendent side groups, or throughout the backbone of the monomer.

These polymerisations are controlled through changes in the initiation conditions, e.g. initiator concentration, initiating light intensity or temperature.

11.3   Hydrogel properties

Some properties of hydrogels that are important to their design and use as biomaterials include swelling, mechanical properties and degradation. The general mechanical properties of hydrogels are based on theories of both viscoelasticity and rubber elasticity. Hydrogels are typically weak compared to other polymers, due to the high water content, and thus it is important to analyse the mechanical properties of hydrogels to control these properties for specific applications. Experimentally, the mechanical properties of hydrogels can be difficult to measure due to solvent, e.g. water evaporation during the experiment and thus it is necessary to measure the mechanics with the hydrogel immersed in the desired solvent. Both tensile testing and dynamic mechanical analysis (DMA) are of interest for hydrogels, measuring the rubber elastic and viscoelastic behaviours respectively. For tissue engineering, it is commonly desired to match the mechanical properties of the hydrogel to those of the intended tissue. A common way to control the mechanical properties of the hydrogel is to alter the cross-linking density of the polymer. Additionally, hydrogel mechanics are altered by the polymerisation conditions during network formation. For instance, the amount of solvent during the polymerisation can lead to more cyclisation (where the cross-linking agent cycles back into the same kinetic chain rather than as a cross-link) in the network. Changes in the reaction conditions (pH, temperature, light intensity) can also play a role in polymer mechanics.

Hydrogel swelling is directly linked to mechanical properties since it correlates to the composition and to the cross-linking density of the network. Swelling is a measure of the water content in the network and is typically reported as the ratio of either the mass or the volume of the network in the swollen state to that of the dry state. In addition, swelling can play an important role in transport and diffusion through the hydrogel, which is important in maintaining the viability of encapsulated cells and in the release of entrapped molecules for drug delivery. Swelling is also a dynamic process that changes with time if the hydrogel was not formed at the equilibrium swelling ratio or if the structure changes with hydrogel degradation.

Hydrogels degrade through a variety of different processes. Natural hydrogels typically degrade in the presence of an enzyme that cleaves the backbone of the hydrogel. These enzymes are usually found throughout the body and lead to degradation *in vivo*. Many synthetic polymers are designed to degrade hydrolytically (in the presence of water) and thus degrade due to the presence of body fluids. The ability to control hydrogel degradation is important for biomaterials' applications since degradation controls properties such as hydrogel mesh size, which is important in the release of entrapped molecules and the diffusion of extracellular matrix components produced by encapsulated cells. If a hydrogel degrades too quickly, the material is resorbed before sufficient tissue is produced, whereas if a hydrogel degrades too slowly, it can become a barrier to tissue formation. There are also applications such as cell encapsulation for immuno-isolation where degradation may not be desired.

11.4   Types of hydrogels

There have been many types of hydrogels developed for biomaterials' applications. These hydrogels are either natural or modified natural polymers or are synthetically derived. Commonly investigated natural and synthetic hydrogels are found in Table 11.2. A more detailed description of some of the more widely investigated hydrogels follows.

Table 11.2 Examples of hydrogels

Natural polymers	Synthetic polymers
Fibrin	Poly(ethylene glycol)
Collagen and gelatin	Poly(acrylic acid)
Hyaluronic acid	Poly(vinyl alcohol)
Alginate	Polypeptides
Agarose	Polyphosphazene
Chitosan	Poly(hydroxyethyl methacrylate)
Dextran	Poly(NIPAAm)
Chondroitin sulphate	and many combinations of the above
Poly(L-lysine)	

Collagen

Collagen is a natural protein that is found abundantly throughout the body (in tissues such as tendons and ligaments) and has been widely investigated as scaffolding in tissue engineering. Collagen is commonly cross-linked with various techniques, e.g. glutaraldehyde, carbodiimide, photooxidation to improve the physical properties (mechanics and degradation) and thus widen its application as a biomaterial. For example, glutaraldehyde induces cross-linking of the lysyl amino acid residues on the collagen, which reduces both the rate of *in vivo* degradation and the immunogenicity, allowing the use of collagen from various species. One of the benefits of using collagen is its ability to be resorbed and integrated in the body, which may be altered depending on the amount of cross-linking. Collagen matrices are commonly sterilised by gamma irradiation, ethylene oxide treatment or electron beam irradiation.

Hyaluronic acid

Hyaluronic acid (or hyaluronan) is a glycosaminoglycan that is found throughout the body in various tissues and fluids and binds to specific cell surface receptors. Hyaluronic acid is the only non-sulphated glycosaminoglycan that consists of repeating units of N-acetyl-D-glucosamine and D-glucuronic acid and is degraded in the presence of hyaluronidases. Like collagen, hyaluronic acid is typically chemically modified to alter its physical properties. One common technique involves the esterification of hyaluronic acid (termed HYAFF), which reduces the water solubility of the material and slows degradation (controlled by the degree of esterification). Hyaluronic acid hydrogels are readily fabricated as microspheres, sponges and fibres depending on the intended application. Numerous other methods, e.g. photocross-linking, aldehyde cross-linking, carbodiimide cross-linking have been investigated to modify hyaluronic acid to create hydrogels.

Fibrin

Fibrin is a natural polymer that forms by the polymerisation of fibrinogen in the presence of thrombin and $CaCl_2$ (blood clotting). To create a fibrin hydrogel, fibrinogen is isolated from blood plasma using various precipitation techniques, e.g. cryoprecipitation, ammonium sulphate precipitation and degraded by thrombin to produce fibrin monomers that aggregate due to hydrogen bonding and become insoluble through the reaction of thrombin and plasma factor XIIIa. For biomaterials applications, fibrin has been formulated as foams, sheets, particles and as glue for use as adhesives and sealants. Fibrin is widely used for cell encapsulation and for the delivery of various growth factors.

Alginate

Alginate is a natural polysaccharide composed of α-D-mannuronic acid and β-L-guluronic acid that is derived from seaweed. Alginate polymers form gels, i.e. ionic cross-links in the presence of various divalent cations, e.g. Ca^{2+}, Mg^{2+}, by cross-linking the carboxylate groups of the guluronate groups on the polymer backbone. Alginate has also been covalently cross-linked and oxidised in an attempt to optimise the physical properties of alginate hydrogels. A limitation to this approach is the limited degradation of covalently cross-linked alginate gels, since cells do not secrete the necessary enzymes for polymer cleavage. Alginate gels are widely used as cell carriers in tissue engineering and as wound dressings.

Poly(ethylene glycol)

Poly(ethylene glycol) is one of the most widely used synthetic polymers due to the extreme hydrophilicity and biocompatibility of these polymers and their resulting hydrogels. Poly(ethylene glycol) is routinely modified with acrylate or methacrylate groups that allow for hydrogel formation in the presence of initiators by a thermally or photoinitiated polymerization. With the introduction of α-hydroxy acid groups, e.g. lactic acid between the polymer and functional group, hydrolytically degradable hydrogels can be formed that have been investigated for both drug delivery and tissue engineering. The degradation is controlled by the stability of the degradable unit, the number of degradable units, and the network cross-linking density. Due to the process of network formation, precursor solutions can be injected into the body (carrying drugs or cells) and then polymerized *in vivo*.

Poly(N-isopropylacrylamide)

Some polymers, including those formed with poly(N-isopropylacrylamide) or P(NIPAAm), exhibit a lower critical phase transition, which is the point where solutions exhibit a phase transition (liquid below and solid above). At this point, the polymer chains precipitate due to an increase in the favourability of polymer–polymer vs. polymer–water interactions. P(NIPAAm) exhibit a reversible phase transition around 32 °C where the chains go from water soluble to water insoluble and the phase change is altered by the addition of hydrophobic and hydrophilic groups. P(NIPAAm) copolymerised with acrylic acid is being developed as an injectable hydrogel for tissue engineering since an increase in gelation is found after reaching physiological temperatures.

11.5   Hydrogels for tissue engineering applications

Every year, organ loss due to trauma or disease results in significant patient morbidity for millions of patients. While the gold standard for organ replacement is transplantation from both autologous and allogenic tissue sources, donor site morbidity (autologous) and donor shortage (allogenic) remain severe limitations. The field of tissue engineering offers great promise in the engineering of new tissue or organs using a number of different strategies (Chapters 18 and 19). A common approach in tissue engineering is to combine the patient's own cells with a polymer scaffold and to use this to generate new tissue *in vitro* or for *in vivo* cell delivery and tissue formation. Many of the polymers described above are being increasingly explored as hydrogel scaffolds for tissue engineering applications. Table 11.3 lists some of the important properties that are desired for scaffolds designed to encapsulate cells.

Table 11.3 Design criteria for hydrogels as scaffolds for tissue engineering

Physical properties
Gel formation mechanisms and dynamics (to allow minimally invasive delivery)
Mechanical properties (appropriate integrity and strength)
Degradation rates (appropriate to application)
Mass transport properties
Diffusion requirements (appropriate diffusion of nutrients and metabolites)
Biological properties
Biocompatibility
Promotion of cell adhesion, proliferation and differentiation

The mechanisms of gelation for hydrogels have been discussed at the beginning of this chapter. A common feature of several of the hydrogels is the ability for gelation to occur under conditions that are not toxic to the cells. This has allowed cells and molecules to be mixed with a polymer solution and injected *in vivo* at which point gelation can be triggered. This minimally invasive cell delivery and *in vivo* gel formation have been demonstrated for alginate, poly(ethylene glycol) and chitosan amongst others. For such approaches, the dynamics of the gel formation is controlled to allow rapid gelation *in vivo*.

The mechanical properties of the hydrogels can be tailored as described earlier in this chapter. For tissue engineering applications, the hydrogel (that is the 'scaffold') may need to provide a load-bearing and/or a volume maintenance function. Additionally, it may be desirable for the scaffold to transmit mechanical stimuli to the cells as necessary for their differentiation and subsequent tissue development. For some load-bearing applications, e.g. bone defects, the relatively low mechanical properties of hydrogels may prove problematic. The mechanisms of hydrogel degradation including hydrolysis, enzymatic cleavage and dissolution have been described earlier in this chapter, and the rate of degradation will be tailored to particular tissue engineering applications. An interesting recent development for tissue engineering has been the incorporation of enzymatically degradable peptide sequences into synthetic hydrogels to allow a cell-responsive degradation of the scaffold.

The mass transport properties of the hydrogels are very important for tissue engineering, as the scaffolds must allow the appropriate transport of nutrients, metabolites, gases and cells throughout the scaffold. Diffusion of molecules and nutrients into, and waste products out of, the scaffold depends on both the molecule and scaffold material properties and their interactions.

The presence and type of nanopores within the gel structure will depend both on the polymer and the gelation conditions. While the molecular weight and Stokes radii of the diffusing molecules compared to the pores will affect the diffusion rate, diffusion may also be affected by charge or other interactions between the polymer chains and the diffusing species. Diffusion of proteins, e.g. albumin, may not occur as freely within highly cross-linked gels as the diffusion of smaller molecules such as glucose and oxygen with Stokes radii less than 1 nm. The hydrogels may also contain pores large enough for cell migration or may be designed to dissolve and degrade over time or in response to cell signals to create pores into which cells can ingress and proliferate.

In terms of the desirable biological properties of the hydrogels, biocompatibility is an essential requirement. This will extend to the biocompatible conditions of gelation in the case of *in vivo* gelation. For tissue development to occur within the hydrogel scaffold, it may be desirable to promote the migration of cells into the scaffold and their subsequent adherence, proliferation and differentiation, including that of encapsulated cells. Hydrogels formed from natural extracellular matrix (ECM) proteins such as collagen may promote cellular adhesion and proliferation. Hyaluronic acid also plays a role in many cell functions and thus is important in imparting biological recognition as a biomaterial. However, many hydrogels lack cell specific receptors and are inherently hydrophilic, leading to poor adsorption of ECM proteins onto the hydrogel surface and, as a result, limited cell adhesion. The lack of recognition and binding to cell receptors has been overcome by modifying polymers such as alginate, P(NIPAAm) and poly(ethylene glycol) with various adhesive peptide and proteins. A common peptide sequence used in this approach is arginine–glycine–aspartic acid (RGD), which is derived from ECM proteins such as fibronectin and collagen and can specifically bind to cellular receptors. Growth factors or growth factor-derived peptides may also be tethered to the scaffolds or released with degradation to promote cell migration, proliferation or differentiation.

Hydrogels as scaffolds for cell delivery and tissue development have been explored in attempts to engineer a wide range of tissues including cartilage, bone, liver, neurons, muscle and fat. In particular, their use in the generation of cartilage has been significantly explored, since the macromolecular structure of cartilage, consisting of a highly hydrated tissue comprising cells within collagen and glycosaminoglycans (GAG), is similar to that of a hydrogel. To this end, alginate has been studied extensively and found to promote the viability and phenotype of cartilage-specific cells (chondrocytes), allowing them to produce ECM proteins such as collagen type II and GAGs that are consistent with cartilage. In general, hydrogels do not possess the greater mechanical properties exhibited by tissues such as bone and tend to be used in non-load bearing bone tissue engineering. Other applications of hydrogels include the use of collagen for engineering blood vessels and alginate as Schwann cell matrices for nerve grafting.

11.6   Summary

Hydrogels are insoluble water-swollen networks that are being widely investigated for biomedical applications such as drug delivery and tissue engineering. They can be formed through a variety of mechanisms including physical and chemical gelation. Properties of hydrogels that are important to their design and use as biomaterials include swelling, mechanics and degradation. There have been many types of hydrogels developed for biomaterials applications. These hydrogels are either natural, e.g. fibrin, collagen and gelatin, hyaluronic acid, alginate and agarose or modified natural polymers or are synthetically derived, e.g. poly(ethylene glycol), poly(acrylic acid), poly(vinyl alcohol) and polypeptides. Many of these hydrogels are increasingly explored as scaffolds for tissue engineering applications such as for the generation of cartilage, bone, liver, neurons, muscle and fat. For such applications, physical properties, mass transport properties and biological properties of the hydrogel must all be considered and tailored for a particular application.

11.7   Reading list

Atala A. and Lanza R.P., *Methods of Tissue Engineering*, St, Louis, Missouri, Elsevier Science, 2002.

Lanza R.P., Langer R. and Chick W.L., *Principles of Tissue Engineering*, London, Academic Press, 1997.

Ratner B.D., Hoffman A.S., Schoen F.J. and Lemons J.E., *Biomaterials Science*, St Louis, Missouri, Elsevier Science, 1996.

Part III
Applications

12

Repair of skeletal tissues

L A R R Y L. H E N C H
Imperial College London, UK

12.1   Introduction

The objective of this chapter is to describe alternative ways to repair bones. The replacement of bones and joints with prostheses is presented in Chapter 13. The topics included in this chapter are (1) mechanisms and rates of bone repair, (2) fracture fixation devices and (3) bioactive materials as bone graft supplements.

12.2   Mechanisms and rates of bone repair

Bone is able to undergo regeneration and to remodel its micro- and macrostructure. This is accomplished through a delicate balance between an *osteogenic* (bone-forming) and *osteoclastic* (bone-removing) process. Bone can adapt to a new mechanical environment by changing the equilibrium between osteogenesis and osteoclasis. These processes will respond to changes in the static and dynamic stress applied to bone; that is, if more stress than the normal physiological level is applied, the equilibrium tilts toward osteoclastic activity (this is known as Wolff's law of bone remodelling).

Nature provides different types of mechanisms to repair fractures in order to be able to cope with different mechanical environments surrounding a fracture. For example, incomplete fractures (cracks) that only allow micromotion between the fracture fragments heal with, or without, a small amount of fracture-line *callus*. This is known as *primary healing*. In contrast, complete fractures, which are unstable and therefore generate macromotion, heal with a large callus stemming from the sides of the bone. This is known as *secondary healing*.

A summary of the steps involved in the repair of a bone fracture is shown in Fig. 12.1 (CD Fig. 12.1). The drawings show the periosteal collars that form, approach each other and fuse in the repair of a fracture. The drawing also shows the formation of internal callus and how the trabeculae become cemented to the original fragments. Living bone of the original fragments is

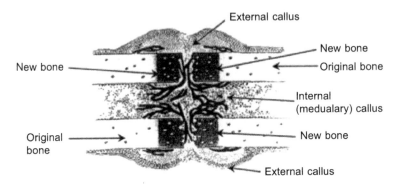

12.1 Long bone repair (CD Fig. 12.1).

light grey and new bone in the external and the internal callus is black. In the external callus, cartilage is stippled lightly and proliferating osteogenic cells are stippled darkly. The time dependence of healing of a long bone fracture is summarised in CD Figs. 12.2 and 12.3.

There are four biomechanical stages of fracture repair, which are summarised in Table 12.1. These correspond to the time-dependent cellular changes shown in CD Fig. 12.2 and variations in type of extracellular matrix in the repair site shown in CD Fig. 12.3. The strength of a healing fracture increases greatly when mineralisation of the osteoid occurs, approximately 4 to 6 weeks after healing starts, as shown in CD Figs. 12.2 and 12.3.

Table 12.1 The four biomechanical stages of fracture repair

Stage	Description
Stage 1	The bone fails through the original fracture site with a low stiffness, rubbery pattern.
Stage 2	The bone fails through the original fracture site with a high stiffness, hard tissue pattern.
Stage 3	The bone fails partially through the original fracture site and partially through the previously intact bone with a high stiffness, hard tissue pattern.
Stage 4	The site of failure is not related to the original fracture site and occurs with a high stiffness pattern.

12.3   Fracture fixation objectives

The goals of fracture treatment are: (1) to obtain rapid healing, (2) to restore function, (3) to preserve cosmesis, and (4) to avoid general or local complications, such as infections. Selection of the treatment method focuses on the need to avoid potentially deleterious conditions, such as excessive motion between bone fragments, which may delay or prevent fracture healing. Each fracture pattern and location, summarised in Chapter 8, CD Fig. 8.35, results in a unique combination of characteristics that require specific treatment methods. The treatments can be non-surgical or surgical. Examples of non-surgical treatments are immobilisation with a cast (plaster or resin) or with a plastic brace. Surgical treatments are divided into (a) external fracture fixation, which does not require opening the fracture site, and (b) internal fracture fixation, which requires opening the fracture site.

With external fracture fixation, the bone fragments are held in alignment by pins placed through the skin onto the skeleton, structurally supported by external bars. With internal fracture fixation, the bone fragment are held by wires, screws, plates, and/or intramedullary devices. CD Fig. 12.4 (a) and (b) show radiographs of externally and internally fixed fractures.

All internal fixation devices should meet general requirements of (a) biocompatibility, (b) sufficient strength within dimensional constraints, and (c) corrosion resistance. The device should also provide a suitable mechanical environment for fracture healing. From this perspective, stainless steel, cobalt–chrome alloys, and titanium alloys are most suitable for internal fixation (see Chapter 2 for details). Most internal fixation devices remain in the body after the fracture has healed, often causing discomfort and requiring removal. Biodegradable polymers, e.g. polylactic acid (PLA) and polyglycolic acid (PGA) (Chapter 10), are used to treat minimally loaded fractures, thereby eliminating the need for a second surgery for implant removal. A summary of the biomaterials used in internal fixation is given in Table 12.2. Compositions and properties of the metallic materials are given in Tables 12.3 and 12.4. Polymeric materials are described in Chapters 4 and 10. Relative success depends on matching the biomaterial properties with device design to meet the biomechanical loads applied to the healing bone. If loads are too great, failure will occur. A description of the principal failure modes of internal fixation devices is presented in CD Table 12.3.

Table 12.2 Biomaterials applications in internal fixation

Materials	Properties	Application
Stainless steel	Low cost fabrication	Surgical wire (annealed); pin, plate, screws, IM nail
Ti alloy	High cost; low density and modulus; excellent osseointegration	Surgical wire, (annealed); pin, plate, screws, IM nail
Co–Cr alloys (wrought)	High cost; high density and modulus; difficult to fabricate	Surgical wire, IM nails
Nylon	Non-resorbable plastic	Cerclage band

12.4   Orthopaedic metals

The three classes of metals most commonly used in orthopaedics are: stainless steel, cobalt–chrome (Co–Cr) alloys and titanium alloys (Table 12.3). 316L stainless steel is resistant to corrosion in saline-rich body fluids due to a high chromium content (17–20% by weight) and low carbon content (less than 0.03%). Addition of molybdenum (Mo) to the alloy improves resistance to pitting corrosion; nickel (Ni) is added to stabilise the austenitic phase at room temperature and enhance corrosion resistance. The mechanical properties of 316L (Table 12.4) depend greatly upon annealing or cold working. The cold worked (wrought) metal is much stronger. Although they are less expensive metals, stainless steels are typically used only for temporary implants, such as discussed in this chapter, due to sensitivity to localised corrosion where stress concentrations and oxygen depletion occurs, as is the case under the screws of a plate used for fracture fixation (see below).

Table 12.3 Composition of metals used in orthopaedics

Element	Stainless Steel 316L (F55–56)	Co–Cr alloys Cast (F 25)	Wrought (F90)	Ti alloy Ti_6Al_4V
Fe	62–69	0.75 max	3.0 max	0.25 max
Ni	12.0–14.0	2.5 max	9.0–11.0	0.1 max
Co	–	62–67	49–56	0.1 max
Cr	17.0–20.0	27.0–30.0	19.0–21.0	0.1 max
Mo	2.0–4.0	5.0–7.0	–	0.1 max
W	–	–	14.0–16.0	0.1 max
Ti	–	–	–	89–91
Al	–	–	–	5.5–6.5
V	–	–	–	3.5–4.5
Mn	2.00 max	1.0 max	2.0 max	0.1 max
P	0.03 max	–	–	0.1 max
S	0.03 max	–	–	0.1 max
Si	0.75 max	1.0 max	1.0 max	0.1 max
C	0.03 max	0.35 max	0.05–0.15	0.08 max
N	–	–	––	0.05 max
H	–	–	––	0.0125 max
O	–	–	––	0.13 max

Two types of cobalt–chromium alloys are used in orthopaedics, one for making products by casting (Co–Cr–Mo: F75) and the other used for forging (wrought) devices (Co–Cr–W–Ni: F90). Tables 12.3 and 12.4 summarise the composition and mechanical properties of both types, for others see the Reading list. The Co–Cr–Mo alloy is very resistant to corrosion in the saline body solutions, even under stress, due to the high Cr content which forms a protective chromic oxide film on the surface. Wrought Co–Cr alloys have

Table 12.4 Mechanical properties of metals used in orthopaedics

	Stainless Steel 316L		Co–Cr alloys		Ti alloy
	Annealed	Cold-worked	Cast (F 25)	Wrought (F90)	Ti_6–Al_4–V
Ultimate tensile strength (MPa)	485	860	655	860	900
Yield strength (0.2% offset) (MPa)	172	690	450	310	830
Elongation (%)	40	12	8	10	10
Modulus of elasticity (GPa)	200	200	220	234	110

excellent ultimate tensile strengths and fatigue lives, which make them the preferred choice for implants that require long service lives without fracture or stress fatigue, such as total joint prostheses (Chapter 13).

The preferred alloy of titanium used for implants is Ti_6–Al_4–V (Table 12.3). The addition of 6% aluminium stabilises the hexagonal close packed (α-Ti) phase and the 4% vanadium addition stabilises a body centred cubic phase (β-Ti) (Chapter 2).

A fine-grained two-phase microstructure results in excellent mechanical strength for the Ti_6–Al_4–V alloy (Table 12.4), with an elastic modulus that is lower than either stainless steel or Co–Cr alloys, which can decrease the stress shielding of bone to a moderate extent. A major advantage of Ti alloys is their superb corrosion resistance due to formation of a very stable TiO_2 oxide phase on the surface during processing. A poor shear strength and fatigue sensitivity to notching makes Ti alloys less desirable for some types of internal fixation devices.

12.5   Fracture fixation devices

Wires

Surgical wires are used to reattach large fragments of bone, such as the greater trochanter, which is often detached during total hip replacement. Wires are also used to provide additional stability in long oblique or spiral fractures of long bones that have already been stabilised by other means. Details are given in the CD module section on wires including CD Fig. 12.5–12.7.

Pins

Straight wires are called Steinmann pins. However, if the pin diameter is less than 2.38 mm, it is called Kirschner wire. These pins are widely used, primarily to hold fragments of bones together provisionally or permanently and to guide large screws during insertion. To facilitate implantation, the pins have different tip designs that have been optimised for different types of bone (CD Fig. 12.8). The trochar tip, which has three cutting faces, is the most efficient in cutting; hence it is often used for cortical bone. The holding power of the pin comes from elastic deformation of surrounding bone. In order to increase the holding power to the bone, threaded pins are used. Most pins are made of 316L stainless steel; however, recently, biodegradable pins made of polylactic or polyglycolic acid have been employed for the treatment of minimally loaded fractures.

Screws

Screws are the most widely used devices for fixation of bone fragments. There are two types of bone screws: (1) cortical bone screws, which have small threads, and, (2) cancellous screws, which have large threads, to get more thread-to-bone contact. Screws may have either V or buttress threads (CD Fig. 12.12). The cortical screws are sub-classified further according to their ability to purchase onto bone: self-tapping (CD Fig. 12.13) and non-self-tapping. The self-tapping screws have cutting flutes, which thread the pilot drill hole during insertion. In contrast, the non-self-tapping screws require a tapped pilot drill hole for insertion. The holding power of screws can be affected by the size of the pilot drill hole, the depth of screw engagement, the outside diameter of the screw and quality of the bone. Therefore, the selection of the screw type should be based on the assessment of the quality of the bone at the time of insertion. Under identical conditions, self-tapping screws provide a slightly greater holding power than non-self-tapping screws. The two principal applications of bone screws are (1) as interfragmentary fixation devices to 'lag' or fasten bone fragments together, or (2) to attach a metallic plate to bone. Details of the variables affecting the performance of screws is given in the CD Chapter 12 section on screws.

Plates

Plates are available in a wide variety of shapes and dimensions and are intended to facilitate fixation of bone fragments. They range from the very rigid, intended to produce primary bone healing, to the relatively flexible, intended to facilitate physiological loading of bone. The underlying goals are (1) to increase the fracture healing rate, (2) to decrease the loss of bone mass in the region shielded by the plate and (3) consequently, to decrease the incidence of refractures which may occur following plate removal. The rigidity and strength of a plate in bending depend on the cross-sectional shape (mostly thickness) and material of which it is made. Consequently, the weakest region in the plate is the screw hole, especially if the screw hole is left empty, due to a reduction of the cross-sectional area in this region. The effect of the material on the rigidity of the plate is established by the elastic modulus of the material for bending and by the shear modulus for twisting. Thus, given the same dimensions, a titanium alloy plate will be less rigid than a stainless steel plate, since the elastic modulus of each alloy is 110 GPa and 200 GPa, respectively (Table 12.4).

Stiff plates often shield the underlying bone from the physiological loads necessary for its healthful existence, termed *stress shielding*. Similarly, flat plates closely applied to the bone prevent blood vessels from nourishing the outer layers of the bone. Use of more flexible plates will allow micromotion and low-contact plates (LCP) allow restoration of vascularity to the bone. The interaction of bone and plate is extremely important, since the two are combined into a composite structure. The stability of the plate–bone composite and the service life of the plate depend upon accurate fracture reduction. Compression between the fracture fragments can be achieved with a special type of plate called a *dynamic compression plate* (CD Fig. 12.16). Details of the clinical use of plates and variables affecting their performance are given in the CD, including CD Figs. 12.16 to 12.24. Excessive bending decreases the service life of the plate. The most common failure modes of a bone plate–screw fixation are screw loosening and plate failure. The latter typically occurs through a screw hole, due to fatigue and/or crevice corrosion (Chapter 2).

Intramedullary nails

Intramedullary devices (IM nails) are used as internal struts to stabilise long bone fractures. Intramedullary nails are also used for fixation of femoral neck or inter-trochanteric bone fractures; however, this application requires the addition of screws. Many designs are available, going from solid to cylindrical, with shapes such as cloverleaf, diamond and C (slotted cylinders). CD Fig. 12.25 shows a variety of intramedullary devices. Compared with plates, IM nails are better positioned to resist multidirectional bending, since they are located in the centre of the bone. However, their torsional resistance is less than that of plates. Therefore, when designing or selecting an intramedullary nail, a high polar moment of inertia is desirable to improve torsional rigidity and strength. The torsional rigidity of an IM nail is proportional to the elastic modulus and to the moment of inertia. For nails with a circular cross-section, torsional stiffness is proportional to the fourth power of the nail's radius. The wall thickness of the nail also affects the stiffness.

In addition to the need to resist bending and torsion, it is vital for an IM nail to have a large contact area with the internal cortex of the bone to permit torsional loads to be transmitted and resisted by shear stress. Two different concepts are used to develop shear stress: (1) a three-point, high-pressure contact, achieved with the insertion of curved pins, and (2) a positive interlocking between the nail and intramedullary canal, to produce a unified structure. Positive interlocking can be enhanced by reaming the intramedullary canal. Reaming permits a larger, longer, nail–bone contact area and allows the use of a larger nail with increased rigidity and strength. Details of the use of IM nails are given in the CD Chapter 12 section on nails. Examples of intramedullary nails are shown in CD Figs. 12.26 to 12.28 A comparison of the extent of callus formation of rod versus plate fixed osteotomies and fractures is given in CD Figs. 12.29 and 12.30.

12.6   Bioactive materials as bone graft supplements

There are many clinical circumstances when bone grafts are required, e.g. replacement of bone following removal of a tumour, revision surgery of failed total hip prostheses, alveolar bone repair after removal of impacted molars, fusion of vertebrae, and repair of chronic bone disunions. Use of an autograft, the patient's own bone, is preferred. However, there is often an insufficient supply of autograft bone available or the quality of bone is poor thus making use of a bone supplement, an allograft, necessary. Bioactive materials are used as allografts, both alone or mixed with small quantities of the patients own bone, often called a pâté, which provides osteoprogenitor cells and bone growth factors, such as bone morphogenic proteins (BMPs).

The concept of bioactive materials having properties intermediate between resorbable and bioinert is reviewed in Chapter 3. A bioactive material is one that elicits a specific biological response at the interface of the material, which results in the formation of a bond between the tissues and the material, shown first in 1969. This concept has now been expanded to include a large number of bioactive materials with a wide range of rates of bonding and thickness of interfacial bonding layers as illustrated in CD Figs. 12.32 and CD Table 12.4. They include bioactive glasses such as Bioglass®; bioactive glass-ceramics such as Cerabone®, A/W glass ceramic or machineable glass-ceramics; dense hydroxyapatite such as Durapatite® or Calcitite® or bioactive composites such as polyethylene–Bioglass®, polysulfone–Bioglass® and polyethylene–hydroxyapatite (HAPEX®) mixtures. All of the above bioactive materials form an interfacial bond with bone. However, the time dependence of bonding, the strength of bond, the mechanism of bonding and the thickness of the bonding zone differ for the various materials as shown in the CD Fig. 12.32. Special compositions of glasses, ceramics, glass-ceramics and composites develop a mechanically strong bond to bone. These compositions are listed in CD Table 12.5.

Some even more specialised compositions of bioactive glasses will bond to soft tissues as well as to bone. A common characteristic of bioactive glasses and bioactive ceramics is a time-dependent, kinetic modification of the surface that occurs upon implantation. The surface forms a biologically active hydroxylcarbonate apatite (HCA) layer, which provides the bonding interface with tissues. The HCA phase that forms on bioactive implants is equivalent chemically and structurally to the mineral phase in bone. It is the biological equivalence of the HCA layer that forms on the bioactive implant surface that is responsible for interfacial bonding. Materials that are bioactive develop an adherent interface with tissues that resists substantial mechanical forces. In many cases the interfacial strength of adhesion is equivalent or greater than the cohesive strength of the implant material or the tissue bonded to the bioactive implant. CD Figs. 12.33 and 12.34 show bioactive implants bonded to bone with adherence at the interface sufficient to resist mechanical fracture. Failure occurs either in the implant (CD Fig. 12.33) or in the bone (CD Fig. 12.34) but almost never at the interface.

Clinical applications of bioactive glasses and glass-ceramics are reviewed in the CD. The successful use of HAPEX®, bioactive PE–HA composite and 45S5 Bioglass® implants in middle ear surgery to replace ossicles damaged by chronic infection is especially encouraging as is the use of A–W glass-ceramic in replacing the iliac crest and in vertebral surgery by Yamamuro. 45S5 Bioglass® implants have been used successfully for maintenance of the alveolar ridge for denture wearers for more than 15 years with nearly a 90% retention rate. The CD section of Chapter 12 describes the clinical use of 45S5 Bioglass® in periodontal repair, maxillofacial reconstruction and orthopaedic repair.

12.7   Summary

Fractured or damaged bone is capable of repair if the bone fragments do not move. Three classes of metals are used in devices to stabilise bone during repair or replacement: stainless steels, cobalt–chrome alloys and Ti alloys. The fracture fixation devices used to immobilise bone are: wires, screws, plates and intermedullary nails.

Bone grafts may be needed to enhance repair in difficult cases. When an insufficient amount of the patient's bone (autograft) is available, synthetic bone graft supplements composed of bioactive glasses or bioactive ceramics are commonly used.

12.8   Reading list

Hench L.L., Bioceramics, *J. Am. Ceram. Soc.* **81** (7), 1705, 1998.
Hughes S.P.F. and McCarthy I.D., *Sciences Basic to Orthopaedics*, Philadelphia, W.B. Saunders, 1998.

Ratner B.D., Hoffman A.S., Schoen F.J. and Lemons J.E. (eds), *Biomaterials Science* (Chapters 2.2, 7.7), Philadelphia, Academic Press, 2004.
Revell P., *Pathology of Bone* (Chapter 9), Berlin, Springer Verlag, 1986.
Simon S.R. (ed), *Orthopaedic Basic Science* (Chapters 7, 9, 10), Illinois, American Academy of Orthopaedic Surgeons, 1994.

13
Joint replacement

LARRY L. HENCH
Imperial College London, UK

13.1 Introduction

Replacement of damaged articular joints with prosthetic implants has brought relief to millions of patients who would otherwise have been severely limited in mobility and doomed to a life in pain. It is estimated that more than 30 million people in the world are affected by osteoarthritis, one of the various conditions that may cause joint degeneration and lead a patient to a total joint replacement. Trauma and fracture related to osteoporosis also lead to a need for joint replacement.

Metallic devices for orthopaedic applications have been very successful with hundreds of thousands being implanted annually. Original applications were as removable devices, such as those for stabilisation of fractures (Chapter 12). Permanent joint replacements began in the 1960s with Professor Charnley's use of self-curing polymethylmethacrylate (PMMA) 'bone cement', which provided a stable mechanical anchor for a metallic prosthesis in its bony bed. This type of anchoring of implants to bone is called 'cement fixation' if PMMA cement is used (Table 13.1). Clinical success of cemented orthopaedic implants has led to rapid growth in use of implants, especially for hip replacements, called total hip arthroplasty (THA) and knee replacements.

The increase in number of implants coincides with an increase in life expectancy of patients and a decrease in average age of patients receiving an implant. This means that a growing proportion of patients will outlive the expected lifetimes of their prostheses. When an implant fails, revision surgery is required. The patient, now 5 to 25 years older, has an increased probability of operative and postoperative complications. This chapter describes the biomaterials and devices used in joint replacements, their survivability and reasons for failure.

Total joint replacements are permanent implants, unlike those used to treat fractures (Chapter 12). The extensive bone and cartilage removed during implantation makes the procedure irreversible. The design of an implant for joint replacement is based on the kinematics and dynamic load transfer

Table 13.1 Applications and properties of common bioinert implant materials

Materials	Applications	Properties
Co–Cr alloy	Stem, head (ball), cup, porous coating	Heavy, hard, stiff, high wear resistance
Ti alloy	Stem, porous coating, metal backing for UHMWPE	Low stiffness
Pure titanium	Porous coating	Excellent osseointegration
Calcium hydroxyapatite	Surface coating	Fast osseointegration, long-term degradation
Alumina	Head, cup	Hard, brittle, high wear resistance
Zirconia	Head	Heavy and high toughness, high wear resistance
UHMWPE	Cup, tibial plateau	
PMMA	Bone cement fixation	Brittle, weak in tension, low fatigue strength

Note: Stem: femoral hip stem/chondylar knee stem; head: femoral head of the hip stem; cup: acetabular cup of the hip.

characteristic of the joint. The material properties, shape, and methods used for fixation of the implant to the patient determine the load transfer characteristics. This is an important element that determines long-term survival of the implant, since bone responds to changes in load transfer with remodelling, e.g. Wolff's law (Chapter 8). Overloading the implant–bone interface or shielding it from load transfer results in bone resorption and subsequent loosening of the implant (stress shielding). The articulating surfaces of the joint must function with minimum friction and produce the least amount of wear products. The implant should be securely fixed to the body as early as possible, ideally during implantation. Removal of the implant should not require destruction of a large amount of surrounding tissues since loss of tissue, especially bone, makes reimplantation difficult and shortens the lifespan of the second joint replacement, called revision surgery.

13.2   Hip joint replacement

The prosthesis for total hip replacement consists of a femoral component and an acetabular component (Fig. 13.1 (CD Fig. 13.1(a)). The femoral stem is divided into head, neck, and shaft. The femoral stem is made of Ti alloy or Co–Cr alloy (less expensive 316L stainless steel is still used in some countries) and is fixed into a reamed medullary canal by cementation or press fitting. Femoral heads are made of Co–Cr alloy, alumina, or zirconia. Although Ti alloy heads function well under clean articulating conditions, they are less used because of their low resistance to third-body wear.

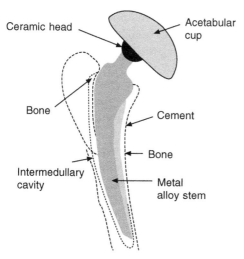

13.1 Schematic of a cemented Charnley hip prosthesis
(CD Fig. 13.1(a)).

Monolithic hip prostheses consist of one part with the advantage of being less expensive and less prone to corrosion or disassembly. Modular devices consist of two or more parts and require assembly during surgery. Modular components allow customising of the implant intra-operatively and during revision surgeries. The length of an extremity can be modified by using a different femoral neck length after the stem has been cemented in place. A worn polyethylene-bearing surface can be exchanged for a new one without removing the metallic part of the prosthesis from the bone. In modular implants, the femoral head is fitted to the femoral neck with a Morse taper, which allows changes in head material and diameter and neck length. Table 13.1, summarises the most frequently used combinations of materials in total hip replacement. Compositions and properties of metals used are given in Tables 12.3 and 12.4.

A monolithic acetabular component is made of ultra-high molecular weight polyethylene (UHMWPE) (Chapters 4 and 10). A modular acetabular component consists of a metallic shell and an UHMWPE insert. The metallic shell minimises microdeformation of the UHMWPE and provides a porous surface for biological fixation of the cup. The metallic shell allows worn polyethylene liners to be exchanged. If a patient has repetitive dislocation of the hip after surgery, the metallic shell allows replacement of the liner with a more constrained one to provide additional stability. The metallic shell and liner must fit precisely because dislodgement of the insert results in dislocation of the hip and damage of the femoral head, since it is in direct contact with the metallic shell. Micromotion between insert and shell produces additional polyethylene debris, which contributes to bone loss.

The hip joint is a ball-and-socket joint. Stability comes from congruity of the implants, pelvic muscles and capsule. Total hip components are optimised to provide a wide range of motion of the prosthetic joint without impingement of the neck of the prosthesis on the rim of the acetabular cup. Designs have evolved to enable implants to support loads that may reach eight times body weight. Proper femoral neck length and correct restoration of the centre of motion and femoral offset decrease the bending stress on the prosthesis–bone interface.

High stress concentrations or stress shielding may result in bone resorption around the implant. The origin of stress shielding is the very large elastic modulus (stiffness) of orthopaedic metals, especially Co–Cr–Mo alloys (10 to 15 times that of cortical bone), as summarised in Chapter 12.

13.3   Failure mechanisms

There are several causes of failure of total joint replacements described in *Clinical Performance of Skeletal Prostheses* (see Reading list) and summarised in Table 13.2.

Table 13.2 Major causes of failure of total joint replacements

Aseptic loosening
(a)   Biological factors including host reaction
(b)   Material properties of the implant component
Infection
Surgical techniques/mechanical causes
Dislocation
Implant fracture
Bone fracture

Aseptic loosening of prosthetic components is the most common cause of failure. Aseptic loosening refers to the failure of joint prostheses without the presence of mechanical cause or infection. It is often associated with osteolysis (bone resorption) and an inflammatory cellular response within the joint. Clinically, aseptic loosening is defined on the basis of the radiographic evidence of the presence of lucent lines at the interface between the bone cement and the implant. Migration of the implant component may also cause significant osteolysis. Osteolysis has also been identified around well-fixed implants. Complications in patients with joint prostheses are often related to the release of particulate wear debris. Particulate material from total joint replacements falls into three categories:

1.   ultra-high molecular weight polyethylene (UHMWPE) from the acetabular component or tibial tray;

2. metal wear debris from acetabular, femoral, or tibial components made from titanium alloy or cobalt chrome alloy;
3. polymethylmethacrylate bone cement.

Load bearing and motion of the prosthesis produces wear debris from the articulating surfaces and from interfaces where there is micromotion. The principal source of wear under normal conditions is the UHMWPE (ultra-high molecular weight polyethylene)-bearing surface in the cup. Many particles are generated with each step, and a large proportion of these particles are smaller than one micrometre in diameter. Cells from the immune system of the patient respond to the polyethylene particles as foreign material and initiate a complex inflammatory response. This response leads to focal bone loss (osteolysis), bone resorption, loosening and/or fracture of the bone. Numerous efforts are under way to modify the material properties of UHMWPE, to harden and improve the surface finish of the femoral head, and to develop other bearing couples, for example, ceramic-to-ceramic and metal-to-metal (Section 13.5).

13.4   Survivability of total hip replacements

The cemented low friction (Charnley-type) total hip arthroplasty (THA), using a metallic femoral component and UHMWPE cup, has the highest level of clinical success, as illustrated in Fig. 13.2. The Kaplan–Meier survivability curve shown in Fig. 13.2 is based upon 42 published clinical studies involving a total of 15 051 patients for time periods as long as 25 years (see *Clinical Performance of Skeletal Prostheses* in the Reading list). The predicted survival rates for the cemented low friction type of THA are:

- 5 years 99.41 ± 0.02%;
- 10 years 95.48 ± 0.04%;
- 15 years 83.12 ± 0.18%;
- 20 years 66.53 ± 0.35%.

A lifetime of 19 years is expected for the average total hip replacement using the cemented low friction prosthesis.

It is uncertain whether the fall-off in survivability between 15 and 20 years, shown in Fig. 13.2, will continue. This is because of the relatively small number of long-term studies reported and the significant improvement in clinical techniques developed during the last 15 years. Advances in surgical procedures should lead to improved 15–25 year lifetimes for cemented THAs.

Life expectancy of THAs depends upon the following factors:

- gender (males lower than females);
- weight (overweight lower than normal);
- age (younger patients with high activity lower than older).

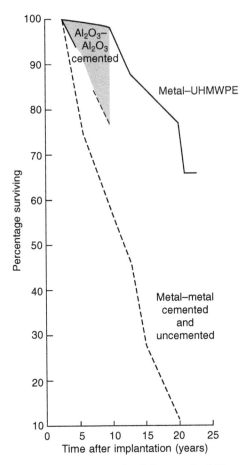

13.2 Schematic graph of the survivability of total hip replacements.

Low friction THA has a low risk of failure for patients > 65 years of age and is generally considered to be the standard for hip replacement.

Other types of THA systems generally show lower survival rates than the cemented low friction type. The alumina–alumina cemented prosthesis is the exception (see below). However, the distribution of results and relatively low number of clinical studies reported for alumina–alumina prostheses currently makes it difficult to establish statistical evidence that this system improves THA survivability by eliminating UHMWPE wear debris.

Uncemented THA systems, regardless of means of fixation, generally show lower survivability results, however, improvements in hydroxyapatite (HA) coatings may lead to longer survivability during the next 10 years. The difference appears by 5 years and expands with time. The difference between cemented and uncemented THA shown in Fig. 13.2 is confirmed by the Havelin *et al.* study of 14 000 patients which shows two times the number of failures of uncemented hips by 5 years (see *Clinical Performance of Skeletal Prostheses* in the Reading list).

13.5   New developments to improve survivability

High-density, high-purity (> 99.5%) Al_2O_3 (alumina) is used as the femoral head in load-bearing hip prostheses for younger patients because of its combination of excellent corrosion resistance, good biocompatibility, high wear resistance and high strength. Most Al_2O_3 devices are very fine grained polycrystalline α-Al_2O_3. Al_2O_3 with an average grain size of <4 µm and >99.7% purity exhibits good flexural strength and excellent compressive strength. These and other physical properties are summarised in Table 13.3 for a commercially available implant material, along with the International Standards Organisation (ISO) requirements. Alumina implants that meet or exceed ISO standards have excellent resistance to dynamic and static fatigue, and resist sub-critical crack growth and implant failure. Load-bearing lifetimes of 30 years to 12 000 N loads or 200 MPa stresses have been predicted and are confirmed by long-term clinical success, with survivabilities of > 80% after 10 years. It is estimated that more than one million hip prostheses to date have been implanted with an alumina ball for the femoral head component and that the number is growing by at least 100 000 per year. Femoral components of total hip prostheses with alumina balls are shown in the CD lecture for this chapter, as discussed above. Clinical studies indicate that alumina/UHMWPE wear rates are approximately ten times lower than Co–Cr–Mo alloy/UHMWPE wear rates of 200 µm per year. Alumina–alumina wear rates are only about 2 µm per year.

Table 13.3 Properties of alumina and zirconia used in surgical implants

Property	Unit	Al_2O_3	TZP	Mg–PSZ
Purity	%	>99.7	97	96.5
Y_2O_3 / MgO	%	<0.3	3 mol	304 wt
Density	g/cm^3	3.98	6.05	5.72
Grain size (average)	µm	3.6	0.2–0.4	0.42
Bending strength	MPa	595	1000	800
Compressive strength	MPa	4250	2000	1850
Young's modulus	GPa	400	150	208
Hardness	HV	2400	1200	1120
Fracture toughness K_{1C}	MPa/m$^{1/2}$	5	7	8

Due to moderate flexural strength and toughness, the size of alumina femoral heads are limited to 32 mm or greater. A higher fracture toughness, flexural strength and lower Young's modulus make medical grade zirconia (ZrO_2) an attractive option for producing smaller femoral heads. A comparison of physical properties of both ceramics is given in Table 13.3. Concerns about long-term survivability of zirconia in the physiological environment compared with 30 years of successful clinical use of alumina makes it difficult to judge the merits of the two materials at this time. A third alternative to either a metal or ceramic femoral head is to produce a coating of ceramic on the surface of a specially designed metal alloy. This is accomplished by converting a zirconium alloy head (97.5% Zr–2.5% Nb) into a ceramic by heating in an oxidising atmosphere to grow a very thin, tough, wear-resistant oxide ceramic layer. The manufacturer reports the oxide coating is nearly 5000 times more resistant to abrasions and scratches than a metallic head and should produce substantially fewer UHMWPE wear debris particles since the coating is 160 times smoother than metal heads.

13.6   Knee joint replacements

The prosthesis for total knee replacement consists of a femoral, a tibial, and/ or a patellar component. Compared with the hip, the knee has a more complicated geometry and movement biomechanics and it is not intrinsically stable. In a normal knee the centre of movement is controlled by the geometry of the ligaments. As the knee moves, the ligaments rotate on their bony attachments and the centre of movement also moves. The eccentric movement of the knee helps distribute the load throughout the entire joint surface.

The prostheses for total knee replacement (see CD Figs. 13.8–13.10) can be divided according to the extent to which they rely on the ligaments for stability. Constrained implants have a hinge articulation, with a fixed axis of rotation, and are indicated when all the ligaments are absent, for example, in reconstructive procedures for tumour surgery. Semi-constrained prostheses control posterior displacement of the tibia on the femur and medial–lateral angulation of the knee but rely on remaining ligaments and joint capsule to provide the rest of the constraint.

Semi-constrained knee prostheses are often used in patients with severe angular deformities of the extremities or in those that require revision surgery, when moderate ligamentous instability has developed. Non-constrained implants provide minimal or no constraint. The prostheses that provide minimal constraint require resection of the posterior cruciate ligament during implantation, and the prosthetic constraint reproduces that normally provided by this ligament.

The prostheses that provide no constraint spare the posterior cruciate ligament. These implants are indicated in patients who have joint degeneration with minimal or no ligamentous instability. As the degree of constraint increases with knee replacements, the need to use femoral and tibial intramedullary extensions of the prosthesis is greater, since the loads normally shared with the ligaments are then transferred to the prosthesis–bone interface.

Table 13.4 Most commonly used knee prostheses

Type I Hinged	Type III Unicompartmental	
Waldius Shiers St Georg Stanmore Guepar	McKeever MacIntosh Porous coated anatomic Marmor Blauth St Georg Sledge Modular Geometric	Oxford Lotus Polycentric/Gunston Huit Unicondylar Robert Brigham Mark I Mark II

Type II Constrained	Type IV Meniscal bearing
Total condylar III Kinematic rotating hinge Sheehan Attenborough Spherocentric	N. J. Low contact stress Accord Oxford Meniscal

Type V Condylar resurfacing

Posterior cruciate ligament retaining	Posterior cruciate ligament sacrificing	Posterior cruciate ligament substituting
Duocondylar Duopatellar Posterior cruciate condylar Kinematic I Kinemax Robert Brigham Townley/anatomic Cloutier AGC Miller–Galante I Anatomic modular knee Press-fit condylar	Total condylar Total condylar II Freeman–swanson ICLH Porous coated anatomic	Insall–Burstein Kinematic stabiliser I Kinematic stabiliser II Kinematic II

Total knee replacements can be implanted with, or without, PMMA cement, the latter relying on porous coating for fixation. The femoral components are typically made of Co–Cr or titanium alloy and the tibial components of UHMWPE. In modular components, the tibial polyethylene component assembles onto a metallic tibial tray. The patellar component is made of UHMWPE, and a metal back is added to components designed for uncemented use.

The wear characteristic of the surface of tibial plateaux differs from acetabular components of total hip replacements. The point contact stress and sliding motion of the knee components may result in delamination and fatigue wear of the UHMWPE. Presumably because of the relatively larger particle size of polyethylene debris, osteolysis around a total knee joint is less frequent than in a total hip replacement. The relatively small size of the patellar component compared with the forces that travel through the extensor mechanism, and the small area of bone available for anchorage of the prosthesis, make the patella vulnerable. The most commonly used knee prostheses are shown in Table 13.4. All total knee arthroplasty (TKA) systems exhibit generally similar survivability results of 82–90% at 10 years.

There is no statistical difference between cemented and uncemented TKAs.

13.7   Ankle joint replacements

Total ankle replacements have not met with as much success as total hip and knee replacements, and they may loosen within a few years of service. This is mainly due to the high load transfer demand over the relatively small ankle surface area and the need to replace three articulating surfaces (tibial, talar and fibular). The joint configurations that have been used are cylindrical, reverse cylindrical and spherical. The materials used to construct ankle joints are usually Co–Cr alloy and UHMWPE. Degeneration of the ankle joint is currently treated with fusion of the joint, since prostheses for total ankle replacement are considered to be still under initial development. CD Fig. 13.13 shows ankle and other total joint replacements.

13.8   Shoulder joint replacements

The prostheses for total shoulder replacement consist of a humeral and a glenoid component. Like the femoral stem, the humeral component can be divided into head, neck and shaft. Variations in the length of the neck result in changes in the length of the extremity. However, since the patient's perception of length of the upper extremity is not as accurate as that of the lower, the various neck lengths are used to fine-tune the tension of the soft tissues, to obtain maximal stability and range of motion.

The shoulder has the largest range of motion in a body, which results from a shallow ball and socket joint that allows a combination of rotation and sliding motions between the joint surfaces. To compensate for the compromise in congruity, the shoulder has an elaborate capsular and ligamentous structure, which provides the basic stabilisation; however, the muscle girdle of the shoulder provides additional dynamic stability. A decrease in the radius of curvature of the implant to compensate for soft-tissue instability will result in a decrease in the range of motion. There is no statistically significant difference between cemented and uncemented shoulder prosthetic systems.

Prior condition of the patient and associated soft tissue defects are the primary factors for success of a shoulder prosthesis. If a shoulder prosthesis is essential, the results show that an unconstrained hemi-arthroplasty or unconstrained total arthroplasty is preferred.

13.9   Elbow joint replacements

The elbow joint is a hinge-type joint allowing mostly flexion and extension but having a polycentric motion. The elbow joint implants are hinged, semi-constrained or unconstrained. These implants, like those of the ankle, have a high failure rate and are not commonly used. The high loosening rate is the result of high rotational moments, limited bone stock for fixation and minimal ligamentous support. In contrast to fusions of the ankle, which function well, fusions of the elbow result in a moderate degree of incapacitation. All types of elbow prostheses provide marked improvement in quality of life, less pain and more range of motion.

Failures of elbow prostheses are uncommon (10–15%) for most types of implants but complications range from 30 to 80%. There are insufficient data to establish statistically significant differences between the types of prostheses used.

13.10   Finger joint replacements

Finger joint replacements are divided into three types: hinge, polycentric and space filler. The most widely used are the space-filler type. These are made of high-performance silicone rubber (polydimethylsiloxane) and are stabilised by a passive fixation method. This method depends on the development of a thin, fibrous membrane between implant and bone. This fixation can provide only minimal rigidity of the joint. Implant wear and cold flow associated with erosive cystic changes of adjacent bone have been reported with silicone implants. Pyrolytic carbon finger joints have been recently developed and show considerable promise for improved clinical success for patients with arthritic joints. See Chapter 14 in *Introduction to Bioceramics* in the Reading list for a description of the structure and properties of pyrolytic carbon.

13.11   Prosthetic intervertebral disc

Fusion of a spinal motion segment in degenerative disc disease increases the stiffness across the stabilised segment and stress on adjacent levels. The results of spinal fusion are unpredictable. It can lead to further degeneration of the adjacent spinal levels. In order to reduce the adverse effects of the fusion process, artificial disc prostheses, similar in concept to the total joint replacements above, have been developed. These designs, although in infancy, range from flexible polymer inserts to ball-and-socket or hinge-type designs.

13.12   Summary

Most of the articulating joints of the body are being replaced routinely by prostheses, the most common being the total hip replacement. Monolithic hip prostheses consist of a femoral stem, typically made from Co–Cr or Ti alloy and an acetabular component made of ultra-high molecular weight polyethylene. Polymethylmethacrylate (PMMA) is used for cement fixation and HA or porous metal coatings are used for uncemented fixation. The most common cause of failure is aseptic loosening caused by bone lysis due to wear debris. Alumina–alumina bearing surfaces minimise wear debris. Survivability of total hip prostheses are excellent; approximately 66% at 20 years. Other joint replacements include knees, shoulders, elbow, fingers and intervertebral discs.

13.13   Key definitions

Bone resorption: a type of bone loss due to the greater osteoclastic activity than osteogenic activity.

Callus: unorganised meshwork of woven bone, which is formed following fracture of bone to achieve an early stability of the fracture.

Fibrous membrane: thin layer of soft tissue that covers an implant to isolate it from the body.

Necrosis: cell death caused by enzymes or heat.

Osseointegration: direct contact of bone tissues to an implant surface without fibrous membrane.

Osteoclastic: activity of bone destruction or removal of old bone by bone cells called osteoclasts.

Osteogenic: activity of bone formation in growth or repair of bone. Bone cells for the osteogenic activity are called osteoblasts.

Primary healing: bone healing in which union occurs directly without forming a callus.

Secondary healing: bone union with a callus formation.

13.14   Reading list

Borozino J.D. (ed), *The Biomedical Engineering Handbook* (Chapter 47), Boca Raton, Florida, CRC Press, 1995.
Havelin L.L., Espohaug B., Vollset S.E. and Engesaster L.B., 'Early failures among 14 009 cemented and 1326 uncemented prostheses for primary coxarthrosis.' *Acta Orthopaedica Scandinavia*, **65**, 1–6.

Hench L.L. and Wilson J. (eds), *An Introduction to Bioceramics*, Singapore, World Scientific, 1993.

Hench L.L. and Wilson J. (eds), *Clinical Performance of Skeletal Prostheses*, Chapman and Hall, 1997.

Hughes S.P.F. and McCarthy I.D., *Sciences Basic to Orthopaedics* (Chapters 16–19), Philadelphia, W.B. Saunders, 1998.

Ratner B.D., Hoffman A.S., Schoen F.J. and Lemons J.E. (eds), *Biomaterials Science* (Chapter 7.7), Philadelphia, Academic Press, 2004.

Simon S.R. (ed), *Orthopaedic Basic Science* (Chapter 9), Illinois, American Academy of Orthopaedic Surgeons, 1994.

14

Artificial organs

J U L I A N R. J O N E S
Imperial College London, UK

14.1 Introduction

Failure of our vital organs leads to death unless a replacement is found. The replacement can either be a transplant from another person or an artificial man-made organ. Use of transplants has the disadvantage of a limited supply and problems with immunorejection. Immunosuppressant drugs are administered to counter this but they have to be taken for life, are expensive and have serious side effects. 'Off-the-shelf' replacement organs offer many advantages, e.g. surgical procedures are quicker, waiting lists can be cut and health services can save money. However, our vital organs are complex and therefore very difficult to mimic using artificial materials; all artificial organs provide only a limited compromise in function.

Our other (non-vital) organs, such as our sensory organs, often need repair, as do the blood vessels that feed them. For the last 30 years, bioinert materials have become used routinely to replace more than 40 different parts of the human body, as illustrated in CD Fig. 14.1. This chapter discusses the status of replacing organs with artificial organs and begins by discussing the replacement of vital organs.

14.2 Kidney

The kidney was the first organ to have an artificial replacement; the dialysis machine invented by Kolff in the 1940s. The kidneys maintain a delicate equilibrium in the blood by controlling the pressure, volume and acidity of the blood, by regulating the concentrations of chemicals, producing hormones and acting as filters. Kidneys can be damaged fatally by birth defects, disease, infection, trauma or toxins. Waste and water removal by the kidneys is too complex a process to mimic. Dialysis is a simultaneous diffusion and filtration process that is different to natural kidney function; it is more simple but very effective. The process is described in detail in Chapter 16.

Dialysis is an effective and life-saving solution but for failing kidneys is not ideal. A dialysis machine will remove urea and unwanted nutrients such as water, sugars and salts from the blood, replacing that function of a natural kidney. However, a natural kidney also re-routes unused nutrients back into the body; a process that dialysis cannot perform. On average, haemodialysis is performed on a patient three times a week for 5 to 6 hours per treatment, costing the patient time and the United Kingdom National Health Service approximately £22 000 per patient per year for hospital dialysis or £19 300 a year if the patient is able to have home dialysis.

David Humes of the University of Michigan is developing a bio-artificial kidney. The device combines a conventional haemo-filtration cartridge with a bioreactor chamber containing a renal tubule assist device containing 10^9 renal proximal tubule (kidney) cells. This bioartificial kidney channels the beneficial elements through tubes lined with the kidney cells, which reabsorb the good nutrients, and sends them through the tubes porous walls into the blood. The device is in early clinical trials, but the patient still needs to attend a clinic for sessions.

14.3   Heart

Heart disease is the largest killer in the USA, killing approximately 1 million Americans every year. The heart suffers badly from the 'western lifestyle' of stress, pollution, excess alcohol and high cholesterol diets. The cardiovascular system was introduced in Chapter 9. The heart's function is to pump blood around the body. The blood contains oxygen and other nutrients vital to all tissues and organs of the body.

The first fully implantable artificial heart was implanted in Robert Tools, a 59-year-old American in Kentucky, July 2001. The 1 kg device, called AbioCor was the size of a grapefruit and is made by Abiomed Inc., Massachusetts. The AbioCor is a titanium and polymer pump, powered by an internal battery and an external battery pack that is worn on the patient's shoulders. No wires pass through the skin, reducing the risk of infection. Mr Tools died within 4 months due to his general ill health and, at the time of writing, 14 patients have received the implant in clinical trials. No causes of death have been attributed directly to the device, and the majority of patients lived longer than their lifetime prediction prior to implantation, as the clinical trials are restricted to patients who are not expected to survive more than thirty days at the time of the operation.

A previous generation artificial heart was the Jarvik-7, which was a four-chamber polyurethane unit with two diaphragm pumps and four mechanical valves. The unit connected quickly into the body but it was only designed to last for 4–5 years and required wires to pass through the skin to an external battery pack. It was in clinical trials in the 1980s but it did not attain long-term success, although it did serve as a replacement heart to keep patients alive for short periods (1 week) while they awaited transplants.

More common strategies to combat critical heart problems are valve replacements, bypass operations (vascular grafts) and heart-assist devices such as pumps and pacemakers, which are described in Chapters 16 and 17. Chapter 17 also discusses artificial blood vessels.

14.4   Lung

The lung exchanges carbon dioxide in the blood for oxygen. Each lung contains tiny air sacs suspended in a net of narrow capillaries that only allow one red blood cell to pass at one time. Each cell excretes carbon dioxide and absorbs oxygen through the sac membranes. The lung contains 40 different types of cells, with a structure too intricate to build artificially, and all its functions are not completely understood. Therefore, at present, only assisted breathing and gas exchange machines have been developed. During surgical operations, blood is removed from the body, and a bubble or membrane oxygenator is used to introduce oxygen into the blood, remove carbon dioxide and return it to the body. Recently, an implantable catheter with an oxygen supply has been developed, which is inserted into the vena cava (vein returning blood to the heart) to assist patients with chronic and advanced lung diseases. As blood passes the catheter, it re-oxygenates. The device can supply an adult with half the oxygen required for a period of up to 2 weeks, so it is useful as a short-term assisted breathing device while the natural lungs recover from trauma or disease.

14.5   Liver

The liver is a large and complex organ, made from cells that have thousands of functions depending on their distance from the arterial supply and varying oxygen contents. The liver is thought to regulate levels of lipids, carbohydrates, blood clotting factors, to regulate the protein metabolism, to synthesise many important chemicals and to detoxify the blood. These functions at present seem impossible to mimic with a wholly artificial construct. At present, the only function that can be mimicked is the detoxification of blood. This is carried out by clinical techniques such as plasmapheresis, which is similar to kidney dialysis, and cryofiltration. In cryofiltration, impurities and pathogens precipitate out of the blood plasma and into a gel as the plasma is cooled. However, both techniques require frequent trips to a clinic to carry out the procedures.

Therefore, the only way to replace a damaged liver is use of an artificial tissue/material hydrid, i.e. a tissue-engineered construct, the principles of which are discussed in Chapters 18 and 19. Alternatively, stem cells may be able to be injected into a damaged liver and regenerate it to its original state and function.

14.6   Pancreas

The pancreas regulates glucose levels in the blood. Clusters (islets) of endocrine pancreatic (β) cells lower glucose levels by increasing insulin hormone levels, opening the glucose receptors in cells. If the pancreas cannot produce insulin, cells cannot adsorb glucose and energy is taken from the break down of fat cells, which causes nausea for the sufferer and may lead to coma and death. Chronic glucose imbalance leads to dehydration, and can lead to failure of sight and of kidneys, heart failure and damage to the circulation of the legs and feet leading to amputation.

In 1969 an implantable insulin pump was developed at the University of Minnesota. It was implanted below the skin of the upper torso and weighed 300 g. Recently a prototype artificial pancreas has been developed by Roman Hovorka at City University, London, that delivers insulin continuously under the skin and maintains blood glucose at a constant level. It consists of three parts: a sensor placed on the skin that measures blood glucose levels, a hand-held computer that analyses this information and a small pump that infuses glucose into the body. Dr Hovorka suggests the device will be small enough for men to fit it on their belts or for women to place it inside their bras and he hopes the product will be on the market in 5 years. The device is functional but not fully implantable and some patients may prefer to use simple insulin injections. It is the sensor part of the device that is causing technological problems.

Ideally, β-cells would be used to produce the correct amount of insulin, acting both as a detector and controlled release device. Therefore, a small implantable biological/artificial hybrid device containing live pancreas cells is an inviting alternative. A major challenge to this approach is keeping islets of insulin-producing pancreas cells alive while protecting them from the body's natural immune system. At the same time, the islet cells must respond to changing glucose levels and release the needed insulin.

Tejal Desai, while a Whitaker Graduate Fellow at Berkeley, built a small capsule employing micromachining techniques, similar to the technology used to make silicon computer chips, which allowed her to etch the pores with nano-sized diameters in a paper-thin silicon membrane. That gave her control over pore number, location and size. The pores were large enough to allow the small-sized glucose, insulin and oxygen to pass through, while blocking immune components, which are larger. iMEDD Inc. (Ohio) have enhanced the initial design, using two silicon wafers glued together with islet cells between, with a port to replenish the cells and a more reliable titanium housing. The new device is about the size of a 50 pence piece. The aim now is to induce capillaries to grow around the device. Improved vascularisation can deliver insulin to the rest of the body faster and increases the transport of nutrients, especially oxygen.

14.7   Skin

The skin is the largest organ in the human body and acts as a barrier, keeping the components of our body in and everything else out, including bacteria. The skin has two layers: the epidermis or outer layer, which is constantly regenerated, and the dermis or inner layer. The dermis does not regenerate but provides mechanical support for the epidermis.

The most common need for artificial skin is a result of burns. Burns can be a physical and social problem. First-degree burns destroy the epidermis. Second-degree burns destroy both layers but leave epidermis around hair follicles, allowing some scar-like regeneration. Third-degree burns destroy all skin, exposing the fat and muscle. Severely burned skin must be removed immediately before bacteria can breed.

The preferred covering for burns is an autograft; however, a patient may not have enough to transplant. Animal or cadaver skin is immediately rejected by the body. The alternative is the use of artificial skin bandages. Op-Site® (Acme United Corp.) is a clear oxygen-permeable polyurethane film that is suitable for first-degree burns. Biobrane® (Woodroof Inc.) is a silicone rubber–nylon compound coated with collagen extract used for more serious burns. These skin bandages can be used for up to 2 months but then a permanent covering is required to reduce scarring.

Skin cells can be harvested and grown on materials in the laboratory to create sheets of skin in a tissue engineering (Chapters 18 and 19) technique; however, this procedure takes time and is therefore not suitable for trauma patients.

An alternative is a two-part artificial skin made from collagen (a protein) and polysaccharide glycosaminoglycan (GAG, a sugar). Bovine collagen is bonded to the GAG, which reduces the rate of resorbtion of the collagen by the body. The two substances are arranged in a precise foam-like architecture to mimic the layers of skin with a silicone top layer. The bonded sheets are wetted and draped over the wound. A neodermis then begins to grow into the inner layer as the graft begins to dissolve. The neodermis is similar to dermis in that it has good mechanical properties and contains nerve cells, but it does not contain the small skin organs like hair follicles and sweat glands. Growth of the neodermis and resorbtion of the artificial skin takes about 3 weeks, after which the surgeon removes the silicone top layer and transplants a postage stamp-sized piece of autograft epidermis, which regenerates to cover the neodermis. The autograft operation can be avoided by adding epidermal cells to the collagen–GAG membrane, before the artificial skin is applied, which will then regenerate epidermis over the neodermis.

14.8   The ear

The ear system converts sound waves to electronic nerve signals for the brain to decode. The hearing system of the ear is made up from four sections:

1. The outer part of the ear (earlobe) and the ear canal funnel sound inwards towards the middle ear.
2. The middle ear contains the eardrum and three tiny bones (ossicles). The ossicles act as levers and amplify and transmit sound vibrations to the opening of the inner ear.
3. The fluid-filled inner ear (cochlea) contains thousands of tiny sound receptors called hair cells. The hair cells sway as the now fluid sound waves hit them in the fluid-filled space.
4. Thousands of little nerve pathways then transmit sound information from the hair cells up to the hearing centre of the brain called the auditory cortex.

Deafness can result from congenital defects, ageing, trauma, infection or diseases such as meningitis. Deafness can be due to deficiencies in any of the four parts of ear and is classified by which part is affected.

Conductive deafness refers to the inability of the outer or middle ear to transmit sound and is often caused by formation of spongy bone in the middle ear that fixes the ossicles (otosclerosis). The condition can be treated with a hearing aid or surgically with a prosthesis (see below).

Sensorineural deafness results from deficiencies in the cochlea, either due to damage to the hair cells (sensory deafness) or by damage to the nerve fibres. Nerve fibre damage cannot be treated, but the hair cells can be bypassed by using a cochlear implant (see below).

The third class of deafness is central deafness, which is a less common affliction resulting from damage to the brainstem or auditory complex that cannot be treated by an implant.

Conductive deafness

'Hamburger. Hot Dog. Ice Cream.' Five ordinary words, but with great significance. In 1984, Dr Gerry Merwin, an ear–nose–throat (ENT) surgeon at the University of Florida, whispered the words into the ear of a young mother, who was expecting her second child. She was deaf and was desperate to be able to hear her newborn baby cry. Her deafness arose from an infection that had dissolved two of the three bones of her middle ear. Under a local anaesthetic to protect the foetus, Dr Merwin had just implanted the world's first Bioglass® device into her middle ear. The implant was designed to conduct sound waves from her eardrum to the cochlea, and thus restore her hearing.

Bioglass®, discovered in 1969, was the first man-made material to bond to living tissues (Chapter 3). Ear surgeons had inserted other types of middle ear implants in patients but they often failed. The materials used were metals (titanium, Teflon, platinum, tantalum) and plastics (Plastipore), selected because they were as inert and non-toxic as possible in the body. When bioinert materials are implanted into the body, a thin layer of scar tissue forms around them that isolates them from the body. For some clinical needs the scar tissue poses no problem but for a middle ear implant scar tissue can be disastrous. Continual motion of the implant can wear a hole in the eardrum and the implant can come out through the hole, permanently damaging the eardrum.

The Bioglass® middle ear implant tested a new concept in repair of the human body, bioactive bonding. The special composition of glass contained the same compounds present in bones and tissue fluids: Na_2O, P_2O_5, CaO and SiO_2 (Chapter 3). Bone cell stimulating ions are released from the glass after implantation and an apatite (calcium phosphate) layer forms on the glass surface. The apatite layer is similar to bone mineral (Chapter 8) and a bioactive bond forms between the implant and the host bone. The theory underlying bioactive biomedical materials was ready for its final test.

Dr Merwin whispered the words. A big smile appeared on the face of the patient and she repeated, 'Hamburger. Hot Dog. Ice Cream!' The Bioglass® middle ear implant worked. 10 years later in a follow-up study it was still working and the mother could hear her 10-year-old child laughing and singing. Thousands of patients have had their hearing restored in the years since with bioactive middle ear implants, including a bioactive composite material HAPEX® (Chapter 5).

However, many deaf patients cannot be helped with middle ear implants. They are deaf because of damage to their inner ear. The hair cells of their cochlea are not able to convert the mechanical vibrations of sound waves to electrical pulses that travel to the brain.

Inner ear

In many cases of deafness, the nerves leading from the hair cells to the brain are fully functional but the hair cells are damaged or lost. A man-made solution to nerve deafness also began with three ordinary words, 'Watson, come here', commanded Alexander Graham Bell in 1876 to his technician in another room. Thomas Watson heard, came and the world changed. Bell's words were converted into electrical signals by a microphone. The electrical pulses travelled along a wire connecting the rooms and were converted by a speaker back into the words heard by Watson. The telephone was invented. Now a 'bionic ear' implant can restore hearing to such an extent that patients can talk on the phone.

A cochlear implant has three sections:

1. a microphone and sound processor attached to the back of the ear lobe, like a normal hearing aid. The sound processor converts sound to digital information and sends the information to a transmitter antenna that is held against the side of the head by a magnet;
2. an implant that is positioned inside the head but adjacent to the antenna receives the transmitted signal and transforms it into an electrical signal that is then sent to the electrode in the inner ear via tiny wires;
3. the electrode array delivers electrical signals through tiny contacts (electrodes) to the hearing nerve and the hearing nerve carries the sound information to the brain, where it is heard.

At present, cochlear implants are expensive and their use is limited. Extensive training is required. The quality of sound stimulated by the electrodes is poor (but always being improved) and overall results are unpredictable. Aural rehabilitation or therapy is key to the successful use of a cochlear implant for many implant recipients. The implant is not a cure for hearing loss, and cannot resolve all contributing clinical components to hearing loss. A new user will need to practise listening to reach maximum performance. For children, rehabilitation is crucial for the development of language and speech skills. There is also concern, especially for children, about the long-term effects of contact of metallic electrodes on nerves, a pathway for migration of infection and metal corrosion products to the brain.

14.9   The eye

Light strikes the cornea (surface), where it is focused through the pupil to a crystalline lens that further refines the focus so that the light rays are shaped into a cone shape at a nodal point in the eye. The iris controls the pupil diameter to control how much light is allowed into the eye. Once the light rays pass the node, they spread out into a mirrored cone as they pass through a gel-like substance and are focussed onto the retina at the back of the inner eye. The retina consists of light-sensitive nerve endings of rods and cones, which send electronic signals through the optic nerves to the visual cortex of the brain, which are two small areas at the lower rear of the brain. If the visual cortex is damaged, sight can be lost even if the eye and optic nerve are fully functional.

The entire process of seeing an object takes 0.01 s, much faster than the processing time of a camera image by the fastest computer.

A few thousand people per year lose their eyesight due to cornea damage by disease or chemical burns. Transplantation is the usual procedure but, if a donor cannot be found, a synthetic 'penetrating' prosthesis can be used. The implant is a cylindrical lens with a Teflon skirt that is inserted through a gap in the cornea. The implant is covered with periosteum (a specialised connective tissue that is a collagen source) and with conjuctiva (the membrane that lines the eyelid).

For 40 years restoration of vision has also been attempted by transmitting electrical signals to the visual cortex of the brain. The effect is to produce bright spots of light (phosphenes) in the mind's eye. The spots are moved to form an image by sending electrical pulses to different electrodes. The image is similar to an electronic scoreboard or a very poor quality television picture. However, concerns for safety of the patients make the restoration of sight by electrical stimulation still experimental. A functional implanted artificial eye is years in the future.

However, sight is restored for more than a million patients annually who have cataracts removed. An intraocular lens (IOL) is inserted in the eye by the surgeon. Light is focussed through the bio-inert polymer lens of the implant (Chapters 3 and 10). The lens is held in place by flexible polymer loops. Because there is little motion inside the eye the IOL is stable and success rates are very high for many years.

Researchers at MIT and the Massachusetts Eye and Ear Infirmary are developing an eye implant that could restore vision to patients suffering from retinal disease, including macular degeneration, an age-related condition that is the leading cause of blindness. The retinal implant will have two silicon microchips, one of which will provide solar power and process an image of the patient's surroundings taken by a tiny camera mounted on glasses worn by the patient. The other will decode picture information and send electric pulses to the retina's receptor cells that forward visual signals to the brain. So, as computer processing power increases and processors become smaller, eventually an artificial eye may be possible.

14.10   The nose

Bionic noses have been developed that convert odour into electronic signals, but they are machines, rather than appendages, which are not necessarily designed for use in the human body but rather as detectors such as quality assurance detectors for the freshness of food or to match batches of chemicals or perfumes. At the sniffing end of the device a vacuum pump pulls in air, which then passes over pairs of thermistors that are coated by an odour-adsorbing material. As the condensation occurs, the temperature rises slightly on the thermistor, which then sends an electronic signal to the computer. The adsorbing materials used are those found in the olfactory epithelium in the nose, i.e. a fatty material, a protein and a carbohydrate.

14.11   The voice box

The larynx is hollow cartilage at the top of the trachea (windpipe). It has three functions. Firstly, it protects the trachea, preventing choking while swallowing. Secondly, it acts as a valve that allows pressure build-up in the lungs enabling us to lift heavy objects. Thirdly, it produces the voice by the vibration of vocal cords. The mouth then shapes the sound into words.

A disease such as cancer may require the voice box to be removed and the trachea diverted to a hole in the neck (a stoma) for breathing. Speech can then only be achieved using muscles in the oesophagus; however only 15% of patients achieve good speech in this way. A T-shaped Blom–Singer prosthesis can be used to help oesophageal speech. The hollow stem of the prosthesis runs from the trachea to the oesophagus via a hole punctured in the back of the trachea. The prosthesis acts as a valve, allowing air into the oesophagus but prevents liquids or solids from entering the trachea. An inverted rubber diaphragm allows air in during inhalation and prevents loss through the stoma during exhalation so that voice produced by the oesophagus is channelled to the mouth.

Electronic larynxes such as the Servox Inton are available. They contain a vibrating electronic sound source that is activated by a push-button control. When pressed firmly against the neck, it transmits sound into the oral cavity. This sound can be moulded into speech as the person mouths words. These devices often have a second button to help intonation.

A future device may have electrodes implanted on the throat muscles, which will be connected to a mini-computer. The electrodes would record muscle signals that precede speech production. The signals would first have to be mapped and interpreted by the developers to program the device.

14.12 Summary

This chapter introduces the use of artificial organs that have been designed to be used as an alternative to transplants, which are always in short supply. However, most of our vital organs are too complex in function and structure to be mimicked by conventional materials and present bioengineering techniques. Ideally, cells in the damaged organ should be stimulated to regenerate the organ to its natural state and function and if the cells in the organ cannot be stimulated, new viable cells should be introduced. An innovative field called tissue engineering is now developing (Chapters 18 to 22), where stem cells from a patient or another source are seeded onto 'scaffolds' made either from artificial biomaterials such as bioceramics or resorbable polymers or natural materials such as collagen. The cells grow on the scaffolds in a bioreactor. The scaffold provides the required three-dimensional shape for the cellular growth and, with time, the material dissolves and the cells form a living tissue. This is then implanted into the patient to replace the diseased or damaged tissues.

The use of artificial mechanical implants during life is now the norm, with four to five million spare parts inserted into people every year. Our desire for an increased life expectancy requires that we use the information contained in each of our unique set of genes to help repair ourselves and research into stem cells to achieve tissue regeneration is a means of achieving this goal. This is now possible in the laboratory – the challenge now to achieve this in patients, paving the way to an entirely new form of medicine.

14.13 Reading list

Cauwels J.M., *The Body Shop; Bionic Revolutions in Medicine*, Missouri, USA, Mosby, 1986.
Fox R.C. and Swazey J.P., *Spare Parts: Organ Replacement in American Society*, Oxford, Oxford University Press, 1992.
Paul J.P., *Biomaterials in Artificial Organs*, Basingstoke, Palgrave Macmillan, 1984.

Websites

http://www.cdc.gov/nccdphp/bb_heartdisease/
http://www.abiomed.com/
http://www.pbs.org/wgbh/nova

15

Mass transport processes in artificial organs

Imperial College London, UK

15.1   Introduction

This chapter describes some of the factors underlying the effective operation
of implants and extracorporeal devices that will be described in Chapters 16
and 17. It is concerned with the physical principles by which materials are
transported, either in normal tissue, or in association with devices. Tissue-
engineered constructs are a special case since the living cells within them
require the establishment of normal exchange processes such as occur in all
other tissues (see Chapters 18 and 19).

Unicellular organisms and the simplest multicellular organisms are able
to survive since diffusion can occur sufficiently rapidly for metabolites to
enter cells at the rate at which they are being used, and waste products to
leave as soon as they are produced. Solutions of the diffusion equations to be
presented indicate that diffusional transport of a very small molecule such as
oxygen in water will take roughly 16 hours to move 1 cm and about 8
minutes to move 1 mm. These rates would be totally inadequate for larger
metabolic species.

Consequently, all large organisms (even plants) have additional mechanisms
for moving fluids to bring metabolites close to the cells. In this context it is
useful to differentiate between pressure-driven 'convective' transport and
'diffusive' transport, which results from intrinsic molecular motion. In strict
thermodynamic terms, these mechanisms are equivalent, since both involve
a minimisation of free energy in a system, but such treatment is beyond the
scope of this book.

15.2   Convective transport

In larger animals, the cardiovascular system (Chapter 9) is the primary
convective transport system. The movement of blood in various vessels is
driven mainly by the pressure generated by the myocardium. Venous return
(the return of de-oxygenated blood through the venules and veins to the

heart) though, like lymph flow and most fluid movement in extracellular spaces, is strongly affected by body movement. The physical mechanisms underlying convective transport are important not only in understanding fluid movement within the body, but also in the design of many extracorporeal devices such as gas exchangers and dialysers.

Interestingly, the simplest law describing fluid flow in tubes was derived in around 1840 while Jean Poiseuille was attempting to understand haemodynamics (although independently around the same time by Gotthilf Hagen). This equation relates the flow of fluid to the pressure required to drive the flow and is expressed as:

$$P_1 - P_2 = \Delta P = \frac{128 \,\mu L Q}{\pi d^4}$$ [15.1]

where P_1 and P_2 are the pressures at the upstream and downstream ends of a tube, ΔP is the pressure difference along the tube, μ is the viscosity of the fluid, L is the length of the tube, Q is the fluid flow rate and d is the tube diameter. This equation can be rewritten in terms of the fluid velocity since, for a tube of circular cross-section, $Q = \bar{U} \pi r^2$ where $\bar{U}$ is the average velocity and r is the tube radius.

Poiseuille's equation can only be applied quantitatively in narrowly confined conditions: (a) the flow must be steady (unchanging in time), (b) the flow must be fully developed (away from entrances, discontinuities, bends) and (c) the tube must be uniform. Such conditions are virtually never met in the body or in engineered systems but the equation is still useful for giving an approximate indication of pressures and flows in many situations. The predicted pressure change ΔP is only applicable to a horizontal tube. Otherwise, the effects of gravity must be taken into account.

From Poisseuille's law we can obtain an estimate of the hydraulic resistance of a system which is given by:

$$\Delta P / Q = \frac{128 \mu L}{\pi d^4}$$ [15.2]

and, for flows through a complex system, series and parallel components may need to be considered.

Features of Poiseuille flow

Fluid 'particles' move in streamlines parallel to the tube walls (axisymmetric flow). These streamlines may be visualised by injection of dye or by neutral density particles, or by using Doppler techniques (ultrasound or laser light) (Fig. 15.1, CD Fig. 15.2).

The distribution of fluid velocities across the tube (the velocity profile) is parabolic and the velocity at any position r is given by:

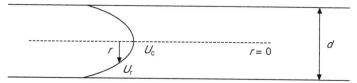

15.1 Schematic of a velocity distribution in a tube (CD Fig. 15.2).

$$U_r = 2\overline{U}\left(1 - \frac{r^2}{d^2/2}\right) \qquad [15.3]$$

Note that the centre line velocity $U_c = 2\overline{U}$ and the velocity at the wall $U_w = 0$. This latter statement is the 'no-slip' condition and presumes an immobile layer firmly associated with the wall.

This description underlies the pressure drops that are a feature of fluid flow, since the different streamlines across the tube are moving at different velocities, and so shearing forces are exerted between the molecules in the different streamlines. The shear rate, s, is the velocity gradient at any point within a flowing system, dU/dr.

In Poiseuille flow, the shear rate is a linear function of the radius:

$$\left(\frac{dU}{dr}\right) = \frac{2U_c}{(d/2)^2} \qquad [15.4]$$

At the axis (centre) of the tube $dU/dr = 0$ and, at the wall, s is maximal and equals $4U_c/d$ or $2\overline{U}/d$.

The shear stress τ, is the product of both shear rate and fluid viscosity: $\tau = \mu \, dU/dr$. A Newtonian fluid is defined as one with constant viscosity, with zero shear rate at zero shear stress, that is, the shear rate is directly proportional to the shear stress. While many fluids approximate closely to having Newtonian properties, this is not the case for blood, mainly because of its cellular composition. Shear stress can be large in many systems and can have adverse effects on blood components, e.g. blood cells can be broken (cell lysis).

It can be seen that fluid components will be sheared much more near the wall than elsewhere in the fluid and since

$$Q = \frac{\overline{U}\pi d^2}{4} \qquad [15.5]$$

then the wall shear

$$\tau_w = \frac{8Q\mu}{\pi d^3} \qquad [15.6]$$

In practice, the shear forces depend strongly on the microscopic geometry of the surface over which the fluid is flowing and it is normally necessary to include an empirical 'roughness factor' for Poiseuille's law to be applicable.

The roughness factor will not only affect the pressure drop along the tube but also the shear forces experienced by the fluid.

Deviations from Poiseuille's Law

Entrance flow

If fluid is moving from an infinite reservoir into a cylindrical tube, at the entrance all the fluid moves at the same velocity and there will be a flat velocity profile. However, the tube surface will retard the fluid (no-slip condition) and viscosity will modify the initial flat velocity profile as more of the fluid is sheared. If the fluid is incompressible, fluid at the centre of the tube must then be accelerated as fluid nearer the tube wall is slowed (law of conservation of mass). The flow will only then become parabolic far from the entrance. Under these circumstances we will have a 'fully developed' flow profile.

To progress from a flat to a parabolic velocity profile, the fluid is subjected to acceleration (positive in the centre and negative towards the walls). Acceleration requires an additional pressure (force/area) above that required to sustain Poiseuille flow. Consequently, the pressure drop and the shear forces in the entrance region will be enhanced.

In the region where viscosity is influencing the flow, shear is occurring and such a region is defined as a boundary layer. In an entrance region, the boundary layer will initially be very thin and, if Poiseuille flow develops, it will come to occupy the whole radius of the tube.

The quantity $U\rho d/\mu$ is dimensionless and is defined as the Reynold's number (Re). It is a measure of the ratio of inertial to viscous forces in the flowing fluid. If $Re > 1$, then inertia dominates the flow patterns and viscosity will only determine the flow near the boundaries, while if $Re < 1$, then flow will be viscous. Not surprisingly, the Re can indicate the distance over which the boundary layer is forming and Poisueuille flow developed from a flat velocity profile is given by $0.03\ Red$, where d is the tube diameter. An entrance region will occur whenever there is a discontinuity in a flowing system, whether it is a change in tube dimensions, a branch or simply a change in direction.

Turbulence

If Re is very high, the inertial forces of the fluid molecules allow breakdown of the fluid streamlines, which are moving parallel to the walls. Then the fluid molecules start moving randomly in micro-eddies causing turbulent flow. This normally occurs above a 'critical Reynold's number' of 2000. Since the micro-eddies carry momentum radially across the tube, turbulence causes blunting of the velocity profiles. The maximum velocity will then be much less than twice the mean velocity. However, the micro-eddies and the very narrow boundary layers will be associated with much more shearing, both within the fluid and at the wall so any detrimental effects on fluid components will be magnified. The shear will also cause a much greater loss of pressure. Consequently, as the pressure driving fluid through a tube is progressively increased, the flow will initially be laminar (Poiseuille) where $\Delta P \propto Q$. This will be followed by a transition zone when the flow oscillates between laminar and turbulent flow and finally a fully developed turbulent region where $\Delta P \propto Q^2$.

Flow in converging and diverging tubes

Whilst in uniform tubes the streamlines will be parallel to the tube walls for low Reynold's number flows, this will not be the case when tubes expand, contract or bend. Under these circumstances, the flow patterns become more complex. It is important to contrast these complex flows from turbulence where the path of individual molecules is unpredictable; as long as Re values are relatively low, the molecules will flow along predictable streamlines even though some will have vectors that are not parallel to the walls.

15.2 Schematic of a converging tube (CD Fig. 15.10).

If a fluid moves from a wider to a narrower tube (Fig. 15.2, CD Fig. 15.10), its velocity will increase and the law of conservation of energy demands that there will be a change in pressure. If we consider two points in the tube (1) upstream and (2) downstream of the constriction, we can assume that the mean velocity is initially U_1 and that it increases to U_2 while the corresponding pressures are P_1 and P_2. If the fluid is incompressible fluid, for conservation of mass: $U_1 A_1 = U_2 A_2$ where A represents the tube area.

If we assume that there is negligible viscous loss, the law of conservation of energy gives:

$$KE_1 + PE_1 = KE_2 + PE_2 \qquad [15.7]$$

which can be expressed as the Bernoulli equation:

$$\tfrac{1}{2}\rho U_1^2 + P_1 = \tfrac{1}{2}\rho U_2^2 + P_2 \qquad [15.8]$$

Since $U_2 > U_1$, the right-hand side (RHS) of the equation is negative and a pressure drop ($P_1 > P_2$) occurs despite the absence of viscous effects. The fall in pressure will be greater than that attributable to the viscous dissipation of energy that will occur in a real fluid and is referred to as the Ventouri effect. This fall could cause tube collapse, entrainment of gas from outside the tube or cavitation of the flowing liquid.

If a tube expands, the same principles can be applied but now $U_2 < U_1$ and so $P_1 < P_2$, that is, in the absence of viscous energy loss, a pressure *increase* will occur in the direction of flow. This is contrary to what could occur in a uniform tube. Taking account of the viscous loss, which will occur in reality, would reduce the pressure increase that occurs, a rise in pressure can still occur if $A_2 >> A_1$.

If the direction of flow changes by less than about 5°, then the fluid streamlines can converge or diverge uniformly. But, if there are sudden changes, more complex flow patterns result. In the case of divergent flow from a narrower to a wider tube, the rising pressure that results from the Bernoulli effect tends to retard the flow. As long as there is a net pressure drop along the whole length of the tube, the fluid inertia in the tube centre will be too great to be much affected by the local pressure change but, near the walls where the inertia is low, the direction of flow can be reversed. Flow reversal will result in an eddy or separation zone close to the walls in which the fluid will continue to rotate. Between the separation zone and the point where fluid flows in the normal downstream direction (the reattachment point), there will be a stagnation point where there is no flow. The degree of separation will increase with increasing flow rate, but turbulent flows will separate less readily than laminar flows.

When fluid passes into a narrower region, the local acceleration of the fluid into the gap will cause a pressure drop, which can lead to separation zones upstream of the narrowing. High pressure losses will occur in the region of separation because of the shear stresses between the forward-moving fluid and the recirculating streams.

A small orifice in an otherwise uniform tube will have separation zones both up and downstream. Under these circumstances, the effective cross-sectional area for flow becomes minimal because of the convergent streamlines – the vena contracta. The emerging flow is consequently in the form of a jet with very high velocities and high local shear stresses. The jet may become turbulent and thereby add to the energy losses.

Flow in curved pipes

When flow enters a bend, the velocity vectors will change (vectors include both magnitude and direction). Consequently, the fluid must be accelerated at an angle to the primary direction of flow. As with changes in tube dimensions, the acceleration must be accompanied by a pressure gradient and, in this case, it will act from the outside of the bend towards the inside. But the faster moving molecules, though experiencing the same pressure gradient as the slower moving ones, will change direction less because of their greater inertia. Consequently, the faster moving fluid will move towards the outer part of the bend, but must be replaced by slower moving (lower inertia) fluid from near the wall (Fig. 15.3, CD Fig. 15.15).

Outer wall

Inner wall

15.3 Schematic of flow at a bend in a tube (CD Fig. 15.15).

These radial components of the flow are referred to as secondary motions and they may persist for a long distance downstream of the bend. Within this entrance region, the radial shear stresses will be dissipated progressively and in the absence of another discontinuity, fully developed flow will again be established. Because of the distortion of the axisymmetric flow towards the outside of the bend, there will be much greater shear there than in straight tubes.

Corners will cause more flow separation and the adverse pressure gradients can cause flow separation. We will see that secondary flows can play an important role in artificial organs that involve heat and mass exchangers.

Steady and unsteady flows

With converging and diverging flow through constrictions, fluid velocity will change by what is referred to as convective acceleration. However, the flow does not change with time at any particular point along the tube and consequently it is said to be steady.

If the pressure gradient changes with time, the fluid will experience local acceleration giving rise to unsteady flows. This is the case with the heart and most other pumps. Changes in pressure with time can be expressed in terms of the angular frequency of oscillation, ω. In oscillatory flow, the effects of viscosity on flow patterns depends on the period of oscillation and the boundary layer thickness in these cases can be given by $(\mu/\rho\omega)^{1/2}$. The fraction of the tube diameter occupied by the boundary layer is:

$$\frac{(\mu/\rho\omega)^{1/2}}{r} \qquad [15.9]$$

and the inverse of this expression is defined as α, the frequency (Wormesley) parameter. This parameter indicates the relative importance of inertial and viscous forces in determining fluid motion during oscillatory flow and when α is low (negligible inertia), then flow will be in phase with the changing pressure gradient, but when α is high, then fluid movement will occur out of phase with the pressure gradient and more complex flow patterns will emerge. Fluid nearest to the wall, with least inertia, reverses in flow direction before that at the axis.

15.3   Diffusional transport

Diffusion is a manifestation of the intrinsic motion of molecules (Brownian motion), which occurs in all materials at temperatures above zero degrees absolute. When molecular motion is rapid (as in a gas), diffusion will occur faster than when the molecules are constrained (as in a solid). It should be noted that both diffusion and heat transport depend on Brownian motion and, indeed, the equations used to describe them are analogous.

Expressed in thermodynamic terms, 'passive' diffusion occurs when a system is moving towards a lower free energy state. 'Active' diffusion can only occur when there is an input of energy to bring about movement (e.g. the conversion of ATP to ADP in cells). In general, the major force for passive diffusion is a decrease in free energy by increasing the entropy (randomness) of the system and this will often involve the transfer of material from a region of higher concentration to one of lower concentration.

When diffusion in a system is occurring at a constant rate, it is said to be in a steady state and can be expressed using Fick's first law:

$$J = DA \frac{da}{dx} \qquad [15.10]$$

where J is the flux (mass/time), D is the diffusion coefficient, A is the area across which diffusion occurs, a is the activity of the diffusing material and x is the distance along which diffusion occurs. The activity is defined as $a = \gamma C$ where γ is the activity coefficient and C is the concentration (mass/volume). For 'ideal' solutions, $\gamma = 1$ and the assumption of ideality is commonly made for aqueous solutions.

Consequently, a reasonable approximation to the Fick equation is given by:

$$J = DA \frac{\Delta C}{x} \qquad [15.11]$$

where ΔC is the concentration difference across distance x. The magnitude of diffusion coefficients is given by the Stokes–Einstein equation:

$$D = \frac{kT}{6\pi r\mu} \qquad [15.12]$$

where k is the Boltzmann coefficient, T is the absolute temperature, r is the molecular radius and μ is the viscosity of the medium through which diffusion is occurring. For gas diffusing in a gas phase, $D \sim 2 \times 10^{-5}$ m^2 s^{-1} but, for gases or small molecules diffusing in water, $D \sim 2 \times 10^{-9}$ m^2 s^{-1} due to the higher viscosity. Proteins in water have an even lower diffusion coefficient, $D \sim 10^{-11}$–10^{-12} m^2 s^{-1} due to their large size. Diffusion coefficients in body tissues will be even lower because of the gel-like nature of fluid.

Many systems are composed of multiple components, e.g. membranes, cytoplasm, extracellular matrix, etc. Under these circumstances it is necessary to take into account not only the diffusion coefficients of the diffusing molecule in each component but also the solubilities. For rapid transport it is necessary to have both a high diffusion coefficient and a high solubility. For such systems, the concentrations of the diffusant in each phase are no longer appropriate for incorporation in Fick's equation but can be modified as C/α, where α is the solubility coefficient (amount of solute dissolved per unit volume of solvent). For example, when oxygen diffuses into a cell, the concentration in the cell membrane is higher than that in either the extra- or intracellular fluids since it is about five times more soluble in lipids than in water. Conversely, when a polar solute diffuses into the cell, the concentration will be much lower in the membrane than either side of it. In general, we can assume that, if a solute is excluded from a material, its activity may be high even when its concentration is very low.

When considering the diffusion of gases, it is conventional to use the partial pressure as a measure of their activity. The 'partial pressure' in a solution, i, is the quantity that is in equilibrium with a gas phase containing that gas species at a partial pressure i.

For example, if the arterial O_2 level is said to be 100 mm Hg, the blood would not gain or lose O_2 if it was in contact with gas containing that partial pressure.

Permeability coefficients

When considering the diffusion of materials through complex systems such as membranes or tissues, it is often difficult to define the solubilities and the dimensions (diffusion distances, etc.) of each of the components. To overcome these problems, it is convenient to define a permeability coefficient, P using the equation:

$$J = PA\,\Delta C \qquad [15.13]$$

where ΔC is the concentration difference across the whole system. If we have components in series:

$$1/P_{total} = 1/P_1 + 1/P_2 + 1/P_3 \ldots, \tag{15.14}$$

where P_1, P_2 etc. are the permeabilities of each of the layers, whilst for components in parallel:

$$P_{total} = P_1 + P_2 + P_3 \ldots \tag{15.15}$$

Many systems have porous membranes, for example hollow fibres used in exchangers or biological systems with intercell clefts, fibres, etc. If A_p is the area of the surface occupied by the pores, then the flux is given by:

$$J = DA_P \frac{da}{dx} \tag{15.16}$$

If the pore dimensions, say radius R, are of the same order of magnitude as that of the diffusing molecule, radius r, then the effective area for diffusion is lower than the total pore area and is equal to $\pi(R - r)^2$. Then:

$$\frac{A_{effective}}{A_{geometric}} = \frac{(R - r)^2}{R^2} = \left(1 - \frac{r}{R}\right)^2 \tag{15.17}$$

Osmosis

Osmosis is simply the movement of a solvent down its own activity gradient. However, this motion can generate a pressure if the fluid is moving into an enclosed space and the osmotic pressure (Π) is given by

$$\Pi = RT \sum \Delta C_i \tag{15.18}$$

where R is the gas constant and C_i is the molar concentration of each dissolved solute. In body fluids $\sum C_i = 290$–300 mM (mOsm) and cell lysis can occur when $\sum \Delta C_i$ falls below 250 mM. This is a major reason for the precise homeostasis of fluid and salt concentrations in the body.

Osmosis will only occur if the movement of the solutes is restricted relative to that of the solvent. The degree of restriction to movement of a given solute is expressed using the reflection coefficient (σ) such that $\sigma = 1$ for complete restriction and $\sigma = 0$ for no restriction. Then:

$$\Pi = RT \left[\sigma_1 \Delta C_1 + \sigma_2 \Delta C_2 + \ldots\right] \tag{15.19}$$

Diffusion with reaction

In some cases, a diffusing molecule will undergo a chemical reaction with other substances that it meets. In such situations, the removal of the diffusant by reaction will maintain a high gradient for diffusion. For example, oxygen diffusing into the blood in the lungs or across an oxygenator reacts with haemoglobin, with the result that the concentration of *free* O_2 remains at a very low level. Under these circumstances, the change in concentration with time is given by the diffusional flux plus the reaction rate. If the reaction rate is rapid, the overall rate will be determined by the transport rate. If the reaction rate is slow, then the rate of transport of a material may be determined by the rate at which it is removed by reaction.

Non–steady state diffusion

If the concentration of a molecule is changing with time, then its gradients will change and the diffusion rate will not be constant. This can occur when a new implant is introduced to the body, when a drug is injected, etc. At any site (x) within a medium, the rate of change of concentration with time is given by the excess of material diffusing in, over that diffusing out.

The 'diffusion equation' (in one dimension) is:

$$\left(\frac{\partial C}{\partial t}\right) = \left(\frac{\partial}{\partial x}\left[D\left(\frac{\partial C}{\partial x}\right)\right]\right)$$

[15.20]

As long as D is independent of x, the rate of change in concentration of a material at site x is given by Fick's second law:

$$\frac{dC_x}{dt} = D\frac{d^2C}{dx^2}$$

[15.21]

The solutions to these equations can be complex. An example might be diffusion following the instantaneous injection of a material into a tissue at site $x = 0$ to give a local concentration C_0 at time t_0 (Fig. 15.4, CD Fig. 15.21).

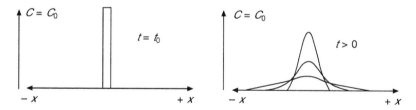

15.4 Schematic graphs showing the solutions to Fick's Law when (a) $t = t_0$ and (b) $t > 0$ (CD Fig. 15.21).

Concentration at site x at time t is given by:

$$\frac{C_{xt}}{C_0} = (\pi Dt)^{1/2} \cdot e^{-x^2/4Dt}$$

[15.22]

Another example might be diffusion from a recently implanted object into the surrounding tissue (Fig. 15.5, CD Fig. 15.22)

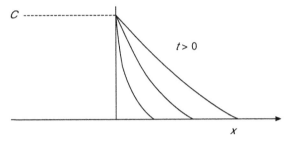

15.5 Schematic graph of a solution to Fick's Law for an object recently implanted into a tissue.

The solution to the equations for this situation shows that the quantity diffusing across interface decreases with $t^{-1/2}$ and that the distance of penetration increases with $t^{1/2}$.

15.4   Interaction of convection and diffusion

In many situations within the body, and in life support systems, the net transport of materials will be by a combination of diffusive and convective processes. When convective and diffusive transport are occurring in the same (or exactly opposite) directions, the total flux can be expressed as the sum: $J_{\text{diffusion}} + J_{\text{convection}}$, though convection is likely to distort diffusion (concentration) gradients, so there will be interaction between these terms.

Ignoring interaction, the flux per unit area of a solute J_s/A, in a flowing system may be expressed as:

$$J_S/A = P\,\Delta C + U_{\text{imposed}}C_0 \qquad [15.23]$$

where P is the permeability coefficient, U_{imposed} is the velocity imposed by convection on the solute, and C is the concentration. For flow through restricted systems, e.g. tissues,

$$U_{\text{imposed}} = U(1 - \sigma) \qquad [15.24]$$

where U is the local fluid velocity and σ is the reflection coefficient introduced above when considering osmosis. With unrestricted diffusion (large pores or small diffusing molecules), then $(1 - \sigma) \to 1$, while for small pores or large molecules $(1 - \sigma) \to 0$. The Peclet number (Pe):

$$Pe = \frac{U(1 - \sigma)d}{D} \qquad [15.25]$$

is a non-dimensional number, which indicates the relative contributions of the two mechanisms, d is a characteristic length; in a tube or pore, it would be the diameter. If $Pe > 1$, the convection dominates, while, if $Pe < 1$, the diffusion is most important.

When the driving forces for diffusion and convection are not in parallel, more complex phenomena result, including the mass transport boundary layer that develops when a material diffuses through a permeable wall into a fluid stream. As it diffuses into the moving fluid, it will be both convected downstream and diffuse radially away from the wall.

On plotting the position reached by the material, we describe the 'mass transport boundary layer' (MTBL), whose thickness may be denoted δ_m. As δ_m increases, the gradient for diffusion between the wall and the fluid will decrease, reducing the rate of mass transport into the fluid, and consequently such layers will limit the efficiency of any dynamic exchanger.

In the steady state, the rate of diffusion into the fluid equals the rate of convection downstream. Mass transport boundary layers will also form when a solute diffuses *out* of a flowing system. In this case, the concentration will be lower in the layer than that in the bulk of the fluid, and outward diffusion will be limited.

Development of these mass transport boundary layers is strongly dependent on the pattern of flow and in particular the nature of the hydrodynamic boundary layer. It can be shown that the ratio of the hydrodynamic to the mass transport boundary layers δ/δ_m, which denotes the Schmidt number (Sc) and is given by the dimensionless quantity $(\mu/D\rho)^{1/2}$, is an indicator of the importance of the MTBL in limiting transport. For transport in the gas phase, $Sc \sim 1$, while for for small molecules in water $Sc \sim 10^3$ and for proteins in water $Sc \sim 10^6$, indicating a decreasing importance of the MTBL in determining transport.

15.5   Dispersion

A material that is present within a flowing stream will be redistributed according to the prevailing flow patterns. For example, if a bolus is injected into a tube in which there is Poiseuille flow, the material at the centre will be convected downstream at twice the average velocity, while that at the wall will be effectively stationary (the no-slip condition). Consequently, the material will be drawn progressively into a very elongated streak, many diameters long.

As a result, there will be a change in the effective concentration along the tube. A similar phenomenon will occur if two fluids are flowing sequentially in tubes; this will occur when one solution is being replaced by another. These are all examples of axial dispersion.

Any diffusion of the material will modify the dispersion. In part, there will be some diffusion in the axial direction, forwards into the stream, but there will also be radial diffusion from faster moving fluid into slower moving fluid and vice versa. Such diffusion will inhibit the axial (longitudinal) dispersion of the material and will blunt the interface.

In Poiseuille flow, the dispersion of a material can be described by an effective diffusivity K, which is given by:

$$K = \frac{(Ud)^2}{192\,D} + D \qquad\qquad [15.26]$$

Deviations from Poiseuille flow, such as secondary flows, cause more radial movement in a flowing fluid and consequently contribute to 'convective mixing' of a material. In turbulent flow, the micro-eddies ensure that there is little axial dispersion. Recirculation zones will also distort dispersion.

15.6   Reading list

Dill K.A. and Bromberg S., *Molecular Driving Forces*, London, Garland Science, 2003.
Fournier R.L., *Basic Transport Phenomena in Biomedical Engineering*, Taylor and Francis, 1998.

Artificial exchange systems

M. JOHN LEVER

Imperial College London, UK

16.1   Introduction

In many areas of medicine various systems are employed either to deliver material to the body or to remove that which is unwanted. Delivery systems include capsules that can be implanted at any location, though commonly subcutaneously (under the skin). These allow the slow release of a drug over a prolonged period of time and eliminate, for example, the need for multiple injections, maintaining a constant level rather than the oscillations that would occur between repeated administrations. The rate of release of the drug can be controlled by its formulation within the capsule and by the properties of the capsule itself. The capsule may be a continuous or porous shell and its thickness and composition can be controlled.

Any implant will initiate a tissue reaction so, in their design, consideration has to be given, for example, to how much fibrous scar tissue may be built up at the interface. These reactions will, of course, have effects on the rate of drug release.

An extension of this mode of therapy is the use of implantable drug delivery pumps that can deliver material remotely through a catheter to, for example, the intrathecal space of the spine for the relief of pain. The advantage of such a system is the possibility of having programmable pumps, which allow the matching of delivery to need. Progress is being made in the application of micro-electromechanical systems (MEMS) in these devices, and in the use of microbionics and miniature power supplies to control release rates.

In other systems, exchange may occur with devices external to the body. These include dialysers, which remove excess fluid and waste products from the blood or fluid and/or toxins from the peritoneal cavity; apheresis devices, used for the collection of fractionated blood components including stem cells; and artificial gas exchangers, which are commonly employed in open-chest surgery and may be used to supplement pathologically ineffective lungs.

For all these devices there must be due consideration of the materials used and particularly those that are in contact with tissues or body fluids. These factors will include sterility, toxicity, resistance to corrosion and other aspects of biocompatibility which are considered in other chapters. There may be particular concerns about the mechanical properties of the components used in pumps, including their strength, compliance and resistance to fatigue.

Dialysers and gas exchangers (as with cardiac assist devices that will be covered in Chapter 17) involve a direct contact between blood and foreign materials. In these cases the material surfaces must be highly resistant to blood clotting. It is essential that devices do not expose the blood to flow conditions that might induce clotting or damage the cells. These include the avoidance of high shear and stagnation regions. To understand these problems, it is important to consider the rheological properties of blood.

16.2   Blood viscosity

Fluid flow is associated with shear between fluid particles in adjacent streamlines. Solid particles in a fluid suspension cannot be sheared readily. Consequently, for a given flow rate, a suspension will exhibit greater shear forces between the fluid particles and hence a higher driving pressure. For these reasons, even a relatively dilute suspension of solid particles will have much higher viscosity than that of the suspending fluid.

Blood cells (particularly red blood cells, erythrocytes) are deformable and so, unlike solid particles, can be sheared themselves and undergo considerable deformation. Blood is consequently very much less viscous than a suspension of solid particles with the same volume fraction (35–55%) and, even after centrifugation, when the packed erythrocytes have a volume fraction of about 97%, they can still flow.

The viscosity of blood though is not constant; blood is a non-Newtonian fluid. In stationary blood the erythrocytes can aggregate to form rouleaux, which branch and form a gel-like structure throughout the whole of its volume. As a result, the blood behaves as a 'Bingham body' and a finite stress (the yield stress) must be applied before it starts to flow (CD Fig. 15.5). At low stresses the fluid has a high viscosity because of the long-chained rouleaux that are present, but on increasing the stress to about 5×10^{-4} Pa, these disaggregate completely and the viscosity reaches a constant value. This shear stress produces a shear rate of about 50 s^{-1}. This is lower than that found in most blood vessels of the body but can occur in distended veins where the blood flow can be very sluggish.

For blood with a normal haematocrit (45%), the viscosity can approach 6×10^{-2} Pa s, at very low shear rates, but this falls to around 3×10^{-3} Pa s at higher levels. This is only about three times greater than the viscosity for plasma and demonstrates how important cell deformability is in allowing efficient blood flow. Not surprisingly though, there is a strong dependence on haematocrit (percentage of red blood cells by volume), and so in individuals with polycythaemia, which is a disorder in which the abnormal bone marrow produces too many red blood cells, the asymptotic viscosity can be greater than that of normal blood by about 50%, and the difference is very much larger at low shear. A major cause of polycythaemia is poor lung function, so unfortunately many patients requiring gas exchange support have the least satisfactory blood properties.

16.3   Effects of shear on blood cells

In addition to concerns about blood viscosity, it is also necessary to consider the effects of shear on blood components. In the healthy circulation, the shear stress even at the blood vessel wall is normally less than 10 Pa, but can be much higher in various disease states or in artificial organs. Erythrocytes, as indicated, are very deformable and bend very easily. This is mainly due to their discoid shape and very large surface:volume ratio. At shear stresses of 1–10 Pa, the cells spin less than at lower shear and tend to align with their long axes parallel to the flow. This alignment actually tends to decrease viscosity. At stress greater than 10 Pa cells are elongated, and above about 160 Pa the cell membranes are stretched and become leaky, first losing ions and later haemoglobin, leading to haemolysis.

Platelets and leukocytes (Chapter 7) are more susceptible to shear damage. Even at shear stresses less than 10 Pa, platelets can leak components such as ADP and 5-HT. These agents can induce aggregation that can lead to coagulation. Leukocytes are the largest blood cells and, consequently, experience the greatest variation of shear across their total surface area. They can release granules and lose their chemotactic properties at shear stresses around 7 Pa. These shear stresses are within the normal physiological range found within the body but, fortunately, damage to cells depends not only on the magnitude of the shear stresses to which they are applied but also the duration. To obtain platelet and leukocyte damage at the levels given above, the shear must be applied for a long period. In many instances in the body, in pumps and in artificial valves, the shear may only be applied for a very short time to any single blood element and so the cells may survive. Some degree of haemolysis and thrombo- and leukopenia are sometimes observed though, particularly during long procedures.

As well as direct effects on blood components, hydrodynamic shear stresses can modify the interaction of proteins, leukocytes or platelets with surfaces. Rudolf Virchow proposed, in the nineteenth century, that thrombosis could depend on a 'triad', i.e. the properties of the blood itself, the surfaces which the blood contact and the flow over those surfaces. The effects on thrombosis of changing the shear can be unpredictable, since moderate levels can inhibit blood–surface interactions, while higher levels can exert prothrombotic effects as described above. To minimise problems associated with the artificial surfaces, it is frequently advantageous to coat solid surfaces with plasma protein, e.g. albumin priming or binding of heparin onto the surface.

16.4　Blood–air interactions

Just as blood components may be adsorbed onto solid surfaces, similar processes occur when blood comes into contact with air. To minimise the free energy in a system, the interfacial tension where two phases make contact is minimised. In relation to blood/gas interfaces, there are two important consequences.

Firstly, proteins will accumulate at the surface, since the intermolecular forces between them and air is greater than between water and air. This adsorption reduces the surface tension; that for water is 0.07 N m^{-1} at body temperature, while for plasma it is in the range 0.04–0.05 N m^{-1}. The conformation of proteins in solution is very strongly determined by the interactions of the molecules with the surrounding water. At an interface, the protein conformation may change, potentially with the alteration of function and sometimes it may invoke an immunological response.

A second consequence of the presence of a liquid/gas interface is that the surface area will be minimised. This is the reason why free bubbles have a minimum surface/volume ratio and are spheres. The surface forces acting circumferentially around a bubble can be resolved into tangential and radial components. The radial component will create a pressure inside the bubble that is in excess of the local hydrostatic pressure and is denoted by $P_{excess} = 2\gamma/R$, where γ is the interfacial (surface) tension and R is the bubble radius. The excess pressure will raise the partial pressure of each of the gaseous components so they can potentially pass into solution in the surrounding liquid. The water vapour pressure will also rise, promoting condensation. Consequently, a spherical bubble is inherently unstable, particularly as the radius decreases.

However, if negative pressures are applied to fluids, they may cause cavitation if the pressure is greater than the vapour pressure of the fluid and dissolved gases and the excess pressure generated by the surface tension. Low pressures must thus be avoided in blood devices, particularly since the proteins reduce the surface tension and facilitate gas separation. Any such bubbles would cause serious problems if they entered the body's circulation. If a bubble enters a small blood vessel such as a capillary, it will no longer be spherical, but will have an excess pressure because of the curvature of the menisci. The pressure is given by $P_{excess} = 2\gamma\cos\theta/r$ where θ is the contact angle between the blood and the wall tissue and r is the vessel radius. Because the capillary radius is very small, this pressure will be very large, higher generally than the pressure generated by the heart. The bubble will then become lodged causing ischaemia of tissue downstream from it.

16.5   Blood flow in artificial devices

The design of a device whether used within the body, such as a valve or pump or an extracorporeal circuit, invariably involves the optimisation of various competing factors. These will include the size of the components that affect the volume of blood, the pressures needed or generated and the shear forces that result from blood flow. For example, small tubes will conserve blood volume but increase the driving forces required and, for a given flow rate, will increase the velocity and hence the shear stresses.

If blood flows in a 1 cm diameter tube at about 3 L/min (lower than the normal cardiac output), the Reynold's number, $Re \sim 2000$. As shown in Chapter 15, this is found to be a critical value, above which streamlined flow will generate into turbulent flow. The greater energy dissipation in turbulent flow will have particularly adverse effects on pressures and flows.

There are also important design concerns about the shape of the spaces through which the blood flows.

Tube narrowing

A connector between tubes, or a tube and a pump, exchanger or filter may decrease the area through which the blood flows. The rise in velocity not only increases the driving pressure needed and the shear applied to the blood but will also be accompanied by a fall in pressure (Bernouilli). This may cause gas entrainment or cavitation.

No-flow regions

External connectors may leave gaps or crevices in which blood will be stagnant. Prolonged contact with a surface invariably will promote thrombosis. Crevices also provide spaces where gas pockets can be stable and provide sites for 'nucleation' of streams of bubbles if the pressure is reduced. When blood enters or leaves a device through a tube, there is invariably a large change in the cross-sectional area through which the blood is flowing. These regions are particularly prone to regions of stagnation.

Bends

Sharp curvature will cause the formation of separation zones with flows at all but the smallest Reynold's numbers. Although the blood is moving in these regions, it remains close to adjacent surfaces for long periods, increasing the risk of both shear damage and thrombosis. However, moderate bends promote the development of secondary flows that can be advantageous in exchangers.

16.6   Exchangers

This section will be concerned with extracorporeal devices that are used to either replace or supplement body exchange processes. It will include discussion of gas and heat exchangers, dialysers and apheresis devices. They are designed to optimise the flux of material between body fluids and other fluids from which they are commonly separated by a membrane.

Diffusion and convection processes must be optimised for efficient exchange. Certain principles apply to all exchangers. These are based on the principles set out in Chapter 15.

1. It follows from Fick's laws that the area for exchange should be as large as practicable. This can be achieved by coiling flat membranes into spirals or by using membranes in the form of hollow fibres; cylindrical bundles of large numbers of these can be arranged as parallel flow components fed and drained by manifolds.
2. The exchange surface must have as high a permeability as possible. The requirements here strongly depend on the properties of the material being transported but in all cases the exchange layer must be thin, and have a high diffusion coefficient and solubility for the diffusing species.
3. The flow of both the blood and the exchange material must be optimised. Channels must be narrow so that all fluid is relatively close to the exchange surface but not so small as to require a larger driving pressure for the fluid flow; this would be associated with increased shear damage. The residence times for which the fluids are in contact with the exchanger must be adequate for a high degree of exchange to occur. To achieve this, the flow rate must be matched appropriately to the volume of the exchange device, and its permeability.

The flow conditions must also limit the development of mass transport boundary layers, which that always hinder efficient exchange. Approaches that have been used to accomplish this have included channels that encourage secondary flows, pulsatile flow and microscopic surface geometries that prevent the development of axial streamlines. In all exchangers some advantage may be achieved by arranging for the blood and exchange fluid to flow in opposite directions (counter-current systems). By this means, the diffusion gradients can be optimised throughout the device and not decline from the inlet to outlet.

Gas exchangers

The modalities used for gas exchange include the use of bubbles and membranes, which may be continuous or porous. Gas exchangers are required during open-chest surgery, particularly when the operation is on the heart itself. They may also be used to supplement cardiorespiratory function when these are severely impaired. It is necessary not only to oxygenate the blood but also to remove excess carbon dioxide. The former is generally easy to achieve since the deoxygenated venous blood can be exposed directly or indirectly to pure oxygen. Assuming the PO_2 of venous blood is typically 50–70 mm Hg, the 'driving force' for oxygen exchange is then typically 650–700 mm Hg. The PCO_2 of venous blood is around 50 mm Hg and, if the blood is exposed to pure oxygen, then the driving force for CO_2 elimination is much lower than that for oxygen. As long as the haemoglobin in the blood is fully oxygenated, it is not important if the PO_2 is not carefully controlled when the blood returns to the body. Conversely, with PCO_2, control is very important, since a level above about 55 mm Hg will cause tissue acidosis, while below about 35 mm Hg will cause alkalosis. In both cases metabolic function will be impaired. For this reason, it is necessary to monitor the pH of the blood re-entering the patient; sometimes it is necessary to incorporate some CO_2 into the exchange gas to avoid alkalosis.

Bubble oxygenators

In these devices, gas exchange occurs directly across the blood/gas film around each bubble. Blood flows up or down a column, through which gas bubbles move upwards under gravity from a sparger in the base. A major advantage of such exchangers is that they present a huge surface area for exchange with minimal boundary layers, since the blood and gas phases are both moving. Although they can achieve higher diffusion gradients for oxygen than for CO_2, CO_2 is much more soluble in plasma and diffuses out of it more rapidly than oxygen diffuses in. Some control over the relative transport rates can be achieved by varying the size of the bubbles, which are typically 30–50 µm in diameter.

   A major disadvantage of bubble oxygenators is that the blood is constantly forming new blood–gas interfaces that can be injurious to both the plasma proteins and the cells. For this reason, bubble oxygenators, although very efficient, are only used for relatively short procedures. Of course, these devices require bubble traps to avoid gas entering the body, and defoaming agents, which increase the plasma surface tension, are needed to prevent froth formation.

Membrane oxygenators

In these devices, the blood is not in direct contact with the gas but is separated from it by a membrane. The membranes must be thin enough (usually 0.2–0.4 μm) to maximise the diffusion gradient and be soluble to gases. For gas exchange, they are invariably made of plastics in which gases are very soluble, such as silicones, polypropylene or siloxanes (Chapters 4 and 10). Because they are so fragile and must sustain the hydrostatic pressure difference between the blood and gas, they need support on a rigid macroporous (pores with diameters in the micrometre range) layer to avoid damage. To achieve adequate exchange, they must have an area of usually 2–5 m^2 and, to accommodate this, the membranes are commonly wound into a spiral with blood and gas spaces of around 1–3 mm in width to minimise shear and resistance to flow. In these exchangers, the ports by which the blood enters and leaves must be designed to avoid stagnant areas and to ensure that the blood is uniformly distributed to the whole exchange surface.

The properties of some materials that have been employed for exchange membranes are given in Table 16.1.

Table 16.1 The exchange properties of certain materials. P_{CO_2}/P_{O_2} is the ratio of permeability of the two gases

Material	Permeability ($D\alpha$)	P_{CO_2}/P_{O_2}
PTFE	8	3
Polyethylene	12	5
Cellulose	25	18
Siloxane	1000	5

A high value of permeability ($D\alpha$) indicates a material that would provide more efficient oxygenation but, for overall gas exchange, account must be taken of the relative permeability.

Porous membrane oxygenators

Polymer membranes can be constructed with multiple pores of 1 μm diameter or less. The membranes can be made from sheets but are most commonly made using hollow fibres. Bundles of these fibres are incorporated into a cylindrical device and blood can flow either through the lumen of the fibres or through the interstices between them, with gas taking the other pathway. Since multiple fibres are perfused in parallel, flow resistance and the shear applied to blood are reduced even though the blood spaces may be as low as 100 μm in diameter. There is some evidence that mass transport boundary layers are reduced if the blood flows through the interstices rather than through the fibres.

Blood is prevented from passing through the pores by surface tension forces. Consequently, the gas is exchanged across a permanent gas/blood interface, which develops as soon as the blood first enters the device and is maintained throughout the procedure. In this way, blood–gas interaction is minimised.

Heat exchangers

During certain surgical procedures it is useful to cool the body to reduce the metabolic rate. To achieve this, the blood is passed through a tube surrounded by a coolant. Typically, the tube takes the form of a coil, not only to reduce the overall dimensions but also to induce secondary motion in the flowing blood and thereby achieve optimal mixing between the blood in contact with the walls and that in the core. Towards the end of the operation the same device is used to rewarm the blood.

Even when cooling is not specifically required, a heat exchanger may be used in other bypass procedures to replace the heat lost as it circulates through the external system.

Filters

It is very important that emboli that are only somewhat larger than blood cells are prevented from passing into the normal circulation of the body. Otherwise, they may block capillaries and cause ischaemia which can be particularly serious if it occurs in the nervous system. Such emboli commonly take the form of small thrombi but may also include gas bubbles, each of which must be removed by filtration.

The filters must be able to filter all material above a certain size and have sufficient capacity so that, if emboli continue to lodge within them over a period of time, there remains sufficient area through which the blood can flow. If not, the pressure upstream can rise excessively and downstream may fall to levels that impede perfusion. Two types of filter are available.

Depth filters are formed from mats of fibres or porous foams. They present a very large area for adsorption of emboli and are not easily blocked. Screen filters, usually constructed of woven polymer fibres, have more precisely determined pores for embolus capture (typically 20–40 μm) but require a large area to prevent blockage. Because filters add to the volume of a bypass system and to the foreign surface to which the blood is exposed, they are not always used and there is some controversy about their use. Filters are particularly important if cardiotomy systems are used. With these, blood that leaks into body cavities during the course of the surgery is sucked up and returned to the patient after filtering.

16.7   Dialysis

Late stage renal failure caused by any one of numerous pathologies can usually only be rectified by dialysis and eventually by kidney transplantation. Dialysers are used to remove excess fluid and waste solutes from the body. They differ from gas exchangers in that they are required to transport water and water-soluble materials. Blood is usually taken from a wrist artery and is passed through a peristaltic pump (pumps are discussed in Chapter 17) to induce a pressure (typically 200–300 mm Hg) sufficient to drive water across an exchanger into a fluid, the dialysate.

The exchangers may be complete or porous membranes. Unlike gas exchangers, they must have a high permeability for water and polar solutes. Cellulose and its derivatives can be used for sheet or hollow fibre. Porous fibres can be made from polysulphones, polycarbonates or polymethylmethacrylate. The sieving characteristics may be varied to allow for different pathologies. For example, patients with massive water retention need large pore haemofilters.

As with other exchangers, the flow conditions must allow adequate time for exchange and must not only minimise mass transport boundary layers, but must also disperse concentration polarisation boundary layers that form by ultrafiltration. Plasma proteins, to which the exchangers are impermeable, will be carried by convection towards the membranes and form a protein-rich layer that will impede dialysis, both by adding an extra layer through which the small molecules must pass (and physically block pores) and by creating an adverse osmotic gradient.

The dialysate must contain appropriate physiological concentrations of ions and nutrients so that there is no net diffusion and loss of these materials from the body. Because the kidney has a role in acid/base homeostasis, the concentration of bicarbonate in the dialysate can be controlled to treat acidosis or alkalosis.

Peritoneal dialysis

If a suitable fluid is pumped through the peritoneal cavity, there can be an exchange of materials between the blood perfusing the abdominal capillaries, particularly the mesenteric vessels and the circulating fluid. Here the exchange membrane is effectively the capillary endothelium across which waste products will diffuse and which are removed from the body in the existing fluid. Use of a hyperosmotic concentration of glucose in the perfusate can be used for extraction of water by osmosis.

The tremendous advantage of this technique is that it can be used in ambulatory patients. A modification of the technique involves the incorporation of chelating agents in the perfusate; this is a common procedure in cases of poisoning by metals such as mercury or cadmium. The reaction of the metal ions and the chelator maintains an optimal diffusion gradient (diffusion with reaction).

Apheresis

This technique is for the separation of components of body fluids (usually blood) either to remove pathological agents from patients or to collect useful material from healthy donors. The two techniques used are either ultrafiltration or continuous centrifugation. Plasmapheresis by ultrafiltration is often used for removing particular toxic proteins such as low density lipoproteins, while leukocytes of circulating stem cells can be collected by centrifugation.

16.8   Summary

This chapter discusses systems that deliver material to the body or that remove what is unwanted. Delivery systems include: implantable capsules for slow, sustained release of a drug and implantable drug delivery pumps that can deliver material through a catheter. Exchange devices that are external to the body include: dialysers, which remove excess fluid and waste products from the blood, apheresis devices that collect fractionated blood components, and artificial gas exchangers, which are commonly used in open-chest surgery and may be used to supplement pathologically ineffective lungs. For all these devices, there must be due consideration of the materials used and particularly for those that are in contact with tissues or body fluids. These factors will include: sterility, toxicity, resistance to corrosion, the mechanical properties of the components used and risk of the materials causing blood clotting. The latter is particularly important for dialysers and gas exchangers as they are in direct contact with blood.

16.9   Key definitions

Acidosis: an abnormal increase in the acidity of the body's fluids caused either by accumulation of acids or by depletion of bicarbonates.

Alkalosis: abnormal increase in the alkalinity of the body's fluids.

Apheresis: procedure in which blood is drawn from a donor and separated into its components, and the remainder returned by transfusion to the donor.

Cardiotomy: a surgical incision of the heart.

Embolus: an air bubble, detached blood clot or a foreign body that travels through the cardiovascular system and lodges so as to obstruct a blood vessel.

Erythrocyte: a red blood cell.

Haematocrit: the ratio of the volume of packed red blood cells to the volume of whole blood as determined by a haematocrit, which is an instrument for determining the relative amounts of plasma and corpuscles in blood.

Hyperosmotic: solution that would cause water loss from body fluids by osmosis.

Ischaemia: localised tissue anemia due to obstruction of the flow of arterial blood.

Mesenteric vessels: blood vessels near a mesentery (any of several folds of the peritoneum that connect the intestines to the abdominal wall).

Peristaltic pump: a pump producing peristaltic movement, which is a series of wave-like contractions by which contents are forced onwards.

Thrombus: a blood clot.

16.10 Reading list

Henrich W.L., *Principles and Practice of Dialysis*, New York, Lippincott, Williams and Wilkins, 2003.

Levy R., Implanted drug delivery systems for control of chronic pain. In *Neurosurgical Management of Pain,* New York, Springer-Verlag, 1997.

Nichols W.W. and O'Rourke M.F., *MacDonald's Blood Flow in Arteries*, London, Arnold, 1998.

Owen W.F., *Dialysis and Tranplantation*, Philadelphia, W.B. Saunders, 1999.

Vogel S., *Vital Circuits*, Oxford University Press, 1992.

17
Cardiovascular assist systems

M. J O H N L E V E R
Imperial College London, UK

17.1   Introduction

The body cannot survive without an intact cardiovascular system that pumps blood around the body and ensures that there is a blood supply to every organ to match its metabolic needs. Consequently, any pathology affecting either the heart or the blood vessels must be repaired if their normal function is inadequate. This chapter will be concerned with various strategies that can be adopted to replace or reinforce different critical components of the system.

A common cause of inadequate cardiac function is valve disease. If valves become stenotic, a large pressure gradient develops across them that may seriously reduce systemic or pulmonary artery pressure and inevitably place a greater load on the heart. Alternatively, valves may fail to close properly, either through age-related distension of the blood vessels or by malformation. Cardiac failure may develop simply through ageing or may be caused by a variety of specific pathologies. When normal pumping can no longer be maintained by therapeutic intervention, there is a need for cardiac assist devices, or artificial or transplanted hearts.

Atherosclerosis is the most common form of vascular disease. Atherosclerotic plaques can impair the circulation either by the formation of stenoses, which can impede blood flow, or emboli, which cause downstream infarction. Plaques may promote the formation of aneurysms which can readily burst, with bleeding into body cavities. Each type of atherosclerosis can be treated by the insertion of autologous or synthetic grafts. More recently though, stenoses are treated preferentially by angioplasty. An angioplasty involves the insertion of a balloon which is inflated to stretch the narrowed vessel and the placement of a stent to maintain patency. Sometimes endarterectomies are performed involving the surgical opening of a vessel with removal of the thickened intimal layer.

All foreign devices implanted into the cardiovascular system increase the incidence of intravascular clotting, which may result in infarction or stroke, therefore minimally thrombotic surfaces are required. Such surfaces may be achieved by selection of the materials used for components, such as chemically binding of a layer of an anticoagulant such as heparin, or, paradoxically, by using roughened surfaces that can form a stable and strongly attached fibrin layer, which inhibits further clotting.

17.2   Heart valves

Valve problems are more common and dangerous when they occur on the left side of the heart. Almost 250 000 repair procedures are performed each year, mainly on the aortic valve but also on the mitral valve. The position of the valves in the heart are shown in Fig. 17.1 (CD Fig. 17.11). Aortic valves are normally repaired with an artificial mechanical or biological prosthesis and, while mitral valves can be replaced, many surgeons prefer to repair the existing valve by reshaping it. Traditionally, valve replacement has required open-chest surgery with a 'heart–lung' machine required to replace cardiorespiratory function, but increasingly, minimally invasive procedures are being developed.

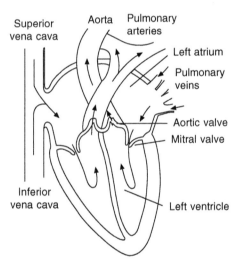

17.1 Schematic of a heart, showing the position of the heart valves and direction of blood flow (CD Fig. 17.11).

For a replacement valve to operate effectively, the following requirements must be met in its design.

Haemodynamics

When fully open, the orifice should be of adequate size to minimise resistance to flow and limit shear. The geometry of the valve should be such as to avoid regions of stagnation or high local stress. The coronary arteries open off the root of the aorta, and prosthetic aortic valves should not impair coronary flow.

Closure mechanism

The movable components should be light to permit full opening with a minimum pressure gradient. They should seal completely, quickly and silently when the pressure gradient is reversed.

Materials

The components should be biocompatible and antithrombogenic. Coagulation around artificial heart valves is the commonest cause of their failure. Strength is an important consideration, and particularly fatigue resistance since, on average, the heart valves open and close about 3.7×10^7 times per year. Although it is desirable for the orifice to be as large as possible, the valve components must be robust enough to avoid fracture, which would inevitably have immediate fatal consequences.

Ease of implantation

Valves are usually attached to the fibrous layer that separates the atria and ventricles by a sewing ring, which is often made of woven polyester. The sewing ring should be of sufficient size for accurate insertion, but not so large as to limit flow.

Mechanical valves

Mechanical valves were first implanted in the 1950s, but their design has been revised continually since then. The solid components of mechanical valves are usually manufactured from stainless steel alloys, molybdenum alloys and increasingly pyrolytic carbon is used for the valve housings and leaflets.

The earliest artificial valves were ball-in-cage devices in which a sphere of plastic or metal is seated, in the closed position, on a metal ring and is retained during opening by a metal cage. With this configuration, blood can no longer flow in a straight path but must be displaced around the ball. These flow conditions increase the work of the heart in propelling the blood. The greater complexity of flow patterns that result inevitably cause more shear damage to the blood components than in devices where the blood flow is not deviated so much. Patients with such valves always need anticoagulation therapy and frequently suffer from haemolytic anaemia.

Better flow characteristics can be achieved with a single tilting disc that lies at approximately 60° to the flow direction when open (Fig. 17.2, CD Fig.17.14). One or two struts are used to constrain the disc, which can be made of plastic or metal. The most recent mechanical valve is a bileaflet valve in which two approximately hemispherical plates pivot on hinges incorporated into a metal ring onto which the leaflets can flatten during closure.

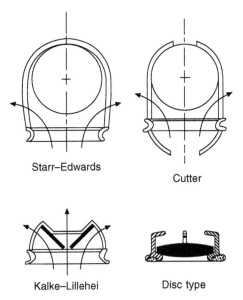

Starr–Edwards

Cutter

Kalke–Lillehei   Disc type

17.2 Schematic examples of mechanical heart valves (CD Fig. 17.14).

Biological valves

Over the past 30 years, various successful approaches have been developed using biological valves. Occasionally, autologous grafts are inserted, when, for example, the pulmonary valve is used to replace a defective aortic valve, with a less critical mechanical or xenograft valve being inserted into the pulmonary artery. Xenografts have been made from porcine or bovine tissue which is 'fixed' with materials such as glutaraldehyde to eliminate immunological reactions. Devices include whole valve systems, single valve leaflets attached to a metal ring, or valves shaped from pericardial tissue. All these valves have a much more normal anatomical geometry than any of the mechanical valves, and coagulation therapy can be relaxed. However, the fixed xenograft tissue may generate problems because of calcification, which makes it stiffer (thus impairing opening and closure) and susceptible fatigue damage, which leads to tearing. The life of xenograft valves is often lower than those of mechnical valves so, despite their rather less satisfactory flow characteristics, modern mechanical heart valves tend to be placed in young patients and may last for their whole lifetime.

Tissue engineering (Chapter 18) has great potential in the area of valve replacement because of the very simple structure of a normal valve leaflet. An ideal example of tissue engineering would be to use autologous vascular cells from a patient's own blood vessels, obtained by biopsy and to grow them on permanent or biodegradable scaffolds to produce valves with no immunological response on implantation. Such valves have been tested in patients at the pulmonary outflow tract.

17.3   Pumps

Mechanical pumps are required most commonly to support the body during procedures such as dialysis or open chest surgery, but are being used increasingly to support the failing circulation in ambulant patients. In the latter role they are usually used to sustain the circulation in the period prior to cardiac transplantation but have occasionally been used for long-term support, working in parallel with the weakened heart.

Cardiac replacement pumps not only have to generate blood flow, but they must also produce appreciable pressures. In dialysers, the pressure is required to drive filtration, and in other extracorporeal devices pumps must overcome the hydraulic resistance imposed by the components of the system. When used inside the body, or when used to return blood to the body, the pressure must be sufficient to drive flow through the microcirculation. While the energy input relative to output is often critical in the design of industrial pumps, in the case of human-support pumps, energy efficiency is very much secondary to safe and reliable operation.

Pumps for extracorporeal devices

Peristaltic (roller) pumps

These are normally used for dialysis and were universally used for heart–lung devices for many years after the technique was introduced in the 1930s. They remain the first choice for many surgeons. In these pumps, two or more rollers rotate on arms through solid semicircular tracks, compressing compliant tubing as they do so. The lumen of the tube may be obliterated effectively so that the roller propels the blood contents in the direction in which it is moving. The rollers are arranged so that, as one comes to the end of the track and releases the tube, another will be compressing the tube at the other end. In this way, the pumps can generate flow against a considerable downstream pressure without the need for valves and may be classified as positive displacement pumps.

The flow rate generated can be predicted fairly accurately from the rate of rotation of the rollers and from the diameter and swept length of the tube. Flow rates are relatively insensitive to the upstream and downstream pressures, although very low downstream pressures can limit efficient re-opening of the tube after it has been compressed. This problem is prevented in practice by using a reservoir above the pump on the inlet side.

A drawback of peristaltic pumps is that very high shear stresses may be imposed on the blood components, particularly during the rapid compression and relaxation of the tube. This problem can be alleviated by using the pump in a non-occlusive mode, in which the tubing is not fully compressed. In this configuration though, the pump will operate less efficiently and will be much more sensitive to upstream and downstream pressures.

Peristaltic pumps produce a weakly oscillatory flow, each peak corresponding to one sweep of a roller. Other pumps, e.g. centrifugal or impeller pumps produce a uniform flow. There is considerable debate about the criticality of this issue. The normal circulation is an oscillatory system and there is some evidence that the physiological control exerted by blood vessel walls depends on pulsatility. Oscillation will also promote mixing within the bloodstream, thereby improving exchange processes which the pumps are intended to support.

Rotary pumps

These may be impellers or centrifugal pumps. To avoid blood leakage, they are normally in the form of encased units with a rotor driven magnetically by an external electric motor. Impellers have rotating blades that force the blood through an enclosed cylindrical space with dimensions little more than the size of the blades. In these pumps very high shearing forces are developed between the blades and the case invariably leads to appreciable haemolysis.

In centrifugal pumps the fluid entering the pump is accelerated rapidly and given a tangential velocity around the axis of a rapidly rotating shaft which may be conical or incorporate vanes. This 'whirl' component not only directs the blood towards the outflow tract but, as it does, it is decelerated since the volume of the casing is increased away from the shaft. This space in which the blood velocity decreases is termed the 'volute'. The deceleration causes a conversion of the high kinetic energy of the blood to potential energy, and this is the mechanism by which these pumps can generate high pressures. With a greater separation of the rotor and the case, compared with the impeller, this design also reduces the high shear stresses applied to the blood.

Centrifugal pumps can have a radial orientation with the blood entering over the axis of the rotor and being displaced to a volute around the perimeter, or they may have an axial orientation in which the blood enters tangentially to the direction of rotation, with the volutes lying on either side of the vanes.

The whirl component, which is the source of the increased pressure, can be estimated by resolving the vectors associated with the velocity of rotation of the blades and the outward velocity of the blood from the rotor into the volute. It follows that as the flow rate of blood through the pump increases, then for a given rotor speed, the whirl component and hence the pressure generated will fall. These considerations dictate the performance characteristics of the pump.

In centrifugal pumps there is no simple way of predicting the flows and pressures produced. These must be derived experimentally and depend strongly on the upstream and downstream pressure conditions. In an ideal situation, it can be shown that there is a linear inverse relationship between flow through the pump and the pressure generated. In practice, a rather lower efficiency is achieved, mainly because of the viscous dissipation by shear forces within the blood. A particular problem arises at the point where the blood rapidly decelerates, since the rising pressure induces separation zones. This problem can be avoided by the incorporation of fixed vanes, which guide the blood in a more controlled manner away from the rotor.

The only means of changing the performance of a given centrifugal pump is to alter the speed of rotation. It can be shown that the pressure generated is proportional to the square of the rotational rate so it is only doubled if the rotation rate is quadrupled. Higher pressures can be generated in pumps with longer vanes, the pressure generated being proportional to the square of the vane length from the axis.

Rotary pumps can produce negative upstream pressures when inflow is restricted, since the rapid acceleration of the blood is accompanied by a sharp fall in pressure (the Bernouilli effect). Such negative pressures can cause cavitation or entrainment of external air.

Pumps for cardiac support

There is an enormous clinical problem relating to heart disease because of the very limited supply of donor hearts for transplantation (Chapter 25). Consequently, there have been many attempts to design artificial hearts. Pumps for the support of ambulatory patients were first introduced in the 1980s, although prototypes had been tested in animals many years earlier. Both moving disc or diaphragm pumps, both of which produce oscillatory flows, and centrifugal or axial flow turbine pumps that produce continuous flows, have been developed. Diaphragm pumps contain a flexible membrane, which is moved backwards and forwards across a chamber into which the fluid flows in and out, through inlet and outlet valves.

Although the early pumps were external to be body, most of those used today are placed in the abdomen. They are supplied by cannulae originating in the right atrium and deliver blood to the ascending or descending thoracic aorta. More recent axial turbine pumps are small enough to be placed within the left ventricle. Modern versions of centrifugal pumps use rotors that are suspended hydrodynamically, thereby minimising the shear that is applied.

Most pumps support left heart function but a few others can support both sides of the heart. It is possible to use a diaphragm pump to operate in reciprocal mode, pumping to the lungs on a forward stroke and systemic circulation as it moves backwards.

Turbine or centrifugal pumps are powered by an electric motor. Disc and diaphragm pumps may incorporate hydraulic or pneumatic drive mechanisms. Most pumps are powered by batteries worn external to the body, but wireless technology now enables transcutaneous power supply to an internal motor or to an internal battery that can be recharged intermittently.

Unlike patients on an extracorporeal bypass where the pump performance can be controlled by a perfusionist, an ambulatory mechanical heart ideally should be responsive to different demands placed on the body. Consequently, most modern implantable hearts are controllable and can change their output, either by input from the patient or ideally by sensors that take signals, for example, from the pacemaker in the right atrium.

The complex three-dimensional structure of the heart presents huge challenges for the production of a complete, tissue-engineered heart. A more promising approach is to use tissue-engineered cardiac myocyte implants to supplement the work of diseased tissue. Progress is also being made in the development of transgenic tissue from pigs and primates.

Ventricular assist devices

When the heart is operating sub-optimally, it is most critical that blood flow to the brain and coronary circulation is sustained. A commonly employed technique that obviates the need for major surgery is to use an aortic balloon pump. A balloon is placed in the thoracic aorta (usually via a femoral artery catheter) and rapidly inflated during diastole and rapidly deflated at the beginning of ventricular ejection. Helium is used for inflation since its low density provides minimal flow resistance during these rapid manoeuvres. By impeding flow to the lower half of the body during diastole, coronary and brain perfusion are enhanced. The rapid deflation aids myocardial contraction by reducing afterload.

Cardiac pacemakers

Cardiac pacemakers are implantable devices that deliver a controlled, rhythmic electric stimulus to the heart to maintain a normal heartbeat. A pacemaker has three components: (1) a pulse generator which contains the battery and circuitry that generates the stimulus and senses electrical activity; (2) an insulated wire lead that carries the stimulus from the generator to the heart and relays intrinsic cardiac signals back to the generator; and (3) a programmable, telemetry device that provides a two-way communication between the generator and the clinician. Advances in materials and electronics have led to long, 10-year, life times for implantable pacemakers. An estimated 350 000 patients worldwide benefit from the improved quality of life that results from use of a cardiac pacemaker.

17.4   Vascular prostheses

If a vessel has becomes stenosed and there is a contra-indication for angioplasty, or if an aneurysm needs major repair, then vascular grafts are inserted to bypass the affected region. Most grafts are currently either autologous or synthetic, although there are many efforts to produce tissue-engineered materials.

A common autologous graft is a coronary artery bypass. In the past, a graft has been harvested from the saphenous vein in the leg and placed between the aorta and the coronary artery, downstream from the stenosis. More recently, a branch of the internal mammary artery is often preferred, firstly because the procedure involves less surgery and secondly because this vessel seems less prone than the saphenous vein to post-operative hyperplasia and re-stenosis.

Synthetic grafts have been fabricated as tubes of woven fabrics including polyesters and polytetrafluoroethylene. Different commercial products frequently incorporate surface layers to resist clotting. Although these prostheses have the strength to resist the considerable mechanical forces that are applied to them, they are not as compliant as the larger arteries that they replace. This lack of compliance presents two problems: firstly, attenuation and reflection of the pressure waves that are an important component of vascular function and secondly, large mismatches in mechanical properties occur at junctions with the native vessels. High local tensile stresses are produced and these promote tissue remodelling leading to damaging hypertrophy of the wall tissue. For these reasons there is great activity in developing elastomeric prostheses, including those incorporating elastin, the main fibrous component of large arteries.

The major advantage of tissue-engineered grafts is that potentially they can be produced with an inner layer of endothelial cells: the cells that line blood vessels in the body, and that have intrinsic antithrombotic and clot-lysing properties. The most promising approaches involve the co-culture of endothelial and smooth muscle cells (the other main cellular constituent of the normal vessel wall) in biodegradable matrices. The smooth muscle cells are able to synthesise the fibrous components of the tissue, which can replace the synthetic matrix as it is broken down.

Stents

Balloon angioplasty has become the preferred method for relieving vascular stenosis, which is caused by atherosclerosis in the majority of patients. Stents are being used increasingly to supplement both angioplasty and endarterectomy and to improve their outcome (Fig. 17.3, CD Fig. 17.15). Both procedures, when performed in isolation, result in a high incidence of re-stenosis and the stents are used to maintain normal vessel dimensions. They are usually cylinders fabricated from metal meshes that are stretched open as angioplasty balloons are inflated. The mesh must be stiff enough to resist the compressive forces imposed on it by the stretched arterial wall tissue and, for this reason, there is particular interest in the use of nickel–titanium 'memory' alloys for these devices (Chapter 2).

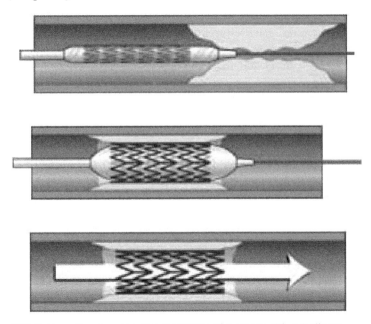

17.3 Schematic showing the insertion of a stent with a balloon angioplasty (CD Fig. 17.15).

Problems that need to be prevented when stents are used include coagulation but also the hyperplastic growth of vessel wall tissue leading to re-stenosis. Various attempts have been made to incorporate growth inhibitors into the stents to prevent these processes.

17.5   Summary

Every organ in the body needs a constant blood supply. Common causes for disruption of the cardiovascular system are heart failure due to age, heart valve disease and vascular disease due to plaque build-up in the blood vessels, which restricts flow or causes aneurysms. Plaque rupture can release emboli or cause thrombosis, blocking the vessel locally or peripherally.

Aged hearts can be assisted by pacemakers, and diseased heart valves can be repair or replaced, either with biological transplants or artificial mechanical prostheses. If none of these solutions is possible, pumps must be used, either extracorporeal or internal artificial hearts, although internal artificial hearts have had limited success. Constricted blood vessels can be widened by angioplasty (insertion of a balloon which is inflated to stretch the narrowed vessel) and by the placement of a stent.

17.6   Key definitions

Anaemia: deficiency in the oxygen-carrying component of the blood.

Aneurysm: pathological, blood-filled dilatation of a blood vessel.

Angioplasty: the surgical repair of a blood vessel.

Atherosclerosis: deposition of plaques containing cholesterol and lipids on the walls of arteries.

Autologous: from the same person's body.

Cannula: a tube used to drain fluid from the body or introduce fluid into it.

Diastole: rhythmical relaxation and dilatation of the ventricles.

Endarterectomy: surgical excision of the inner lining of an artery that is clogged.

Fibrin: a protein that forms the fibrous network in a blood clot.

Heparin: an acidic mucopolysaccharide that has the ability to prevent the clotting of blood, used in the treatment of thrombosis.

Hyperplasia: an abnormal increase in the number of cells in a tissue, causing consequent enlargement.

Hypertrophy: non-tumorous enlargement of a tissue as a result of an increase in the size rather than the number of constituent cells.

Infarction: the obstruction of any organ or vessel of the body.

Re-stenosis: recurrence of stenosis after surgery.

Stenosis: a constriction of a vessel.

Stent: a device that is used to widen blocked blood vessels.

Stroke: sudden loss of brain function caused by damage to a blood vessel to the brain, characterised by loss of muscular control.

Xenograft: transplant from a non-human donor to a human recipient.

17.7   Reading list

Barbenel J., *Blood Flow in Artificial Organs and Cardiovascular Prostheses*, Oxford, Clarendon Press, 1989.

Borozino J.D. (ed), *The Biomedical Engineering Handbook* (Chapter 79), Boca Raton, Florida, CRC Press, 1995.

Goldstein D.J. and Oz M.C., *Cardiac Assist Devices*, New York, Futura Publishing, 2000.

Verdonck P., *Intra and Extracorporeal Cardiovascular Fluid Dynamics*, Volume 1, *General Principles in Application*, Series; Advances in Fluid Mechanics, Vol. 22, UK, Wessex Institute of Technology Press, 1998.

Part IV

Tissue engineering

18

Introduction to tissue engineering

LEE D. K. BUTTERY
University of Nottingham, UK

ANNE E. BISHOP
Imperial College London, UK

18.1 Introduction

Tissue engineering can best be defined by its goal: the design and construction in the laboratory of living, functional components that can be used for the regeneration of malfunctioning tissues. Although considered as being a relatively new field, the first documented report of tissue engineering emerged in 1933, when tumour cells were wrapped in a polymer membrane and implanted into a pig. Tissue engineering is an interdisciplinary field that brings together the principles of the life sciences and medicine with those of engineering and has three basic components: cells, scaffolds and signals. Its development over the past decade has been the result of a variety of factors: increased knowledge and availability of stem cells, genomics, proteomics, the advent of new biomaterials as potential templates for tissue growth, improvements in bioreactor design and increased understanding of healing processes have all contributed. However, although tissue engineering research is evolving rapidly, there has been a hiatus in the commercial development and, hence, clinical application of engineered products. The key challenges to industrial development include problems in devising cost-efficient, scalable processes, guaranteeing product viability and satisfying regulators (Chapter 23). However, progress continues and the number of people currently benefiting from tissue engineering is set to expand exponentially in the coming years. In this chapter, we describe how tissues can be engineered and some of their current applications.

18.2 The challenge

The challenge for tissue engineering is to optimise the isolation, proliferation and differentiation of cells, and to design scaffolds or delivery systems that are conducive to supporting and co-ordinating growth of three-dimensional tissues in the laboratory. One idealistic strategy would be to harvest stem cells from a patient, expand them in cell culture (Chapter 20), and seed them on a scaffold. Stem cells can become many types of specific mature cells, via a process call differentiation, when given the specific biological stimuli. The scaffold should then act as a template and stimulus for proliferation (multiplication) and differentiation of the stem cells into the specific cells that will generate specific new tissue. The tissue can either be grown on a scaffold that will completely disappear (resorb) as the new tissue grows, so that only the new tissue will be implanted, or a 'biocomposite' of the scaffold and new tissue can be implanted (Chapter 19). After implantation, the tissue-engineered construct must then be able to survive, restore normal function, e.g. biochemistry and both mechanical and structural integrity, and integrate with the surrounding tissues. Using cells from the same patient eliminates the problem of immunorejection that can occur with transplants from donors.

18.3   Cell sources

Probably the single most important element in the success of tissue engineering is the ability to generate appropriate numbers of cells (too many cells can be just as detrimental as too few) and the capacity for those cells to differentiate from, and maintain, the correct phenotype and perform specific biological functions. For example, cells must produce an extracellular matrix (a protein-based matrix such as collagen in bone) in the correct organisation, secrete cytokines and other signalling molecules, and interact with neighbouring cells/tissues. Immediately, this raises a number of potential problems, not least of which is obtaining appropriate cell numbers to promote repair.

Tissue engineers have looked at virtually all tissues in the body. In some cases, it has been possible to repair/replace tissue using, as a starting material, the relevant cells from the same patient or from a close relative, such as knee repair using autologous chondrocytes. Non-specific cell types have also been used, including dermal fibroblasts for heart valve engineering. See Chapter 6 for descriptions of cell types and definitions.

Primary cells

Primary cells are mature cells of a specific tissue type that are harvested from explant material removed by surgical procedure. An example is primary human osteoblasts that are harvested from the femoral heads removed during total hip replacement operations. Primary cells are the most desirable with regard to immunological compatibility but, in general, they are differentiated, post-mitotic cells. This means that they are no longer able to divide and their proliferation potential is low. This might be compounded by the tendency of some cell types to de-differentiate during *ex vivo* cultivation and express an inappropriate phenotype, e.g. articular chondrocytes in culture often produce fibrocartilage as opposed to hyaline cartilage. This has stimulated studies to find and develop alternative cell sources for tissue engineering strategies, and stem cells might represent a solution to the limitations of primary cells obtained from explanted tissues.

Stem cells

Stem cells are commonly defined as undifferentiated cells that can proliferate and have the capacity to both self-renew and differentiate to one or more types of specialised cells. However, there has been some reconsideration of this definition recently following the observation of de-differentiation and trans-differentiation of certain mature cell types. In view of this, it has been suggested that it should be made broader and applicable to a biological function that can be induced in a range of cell types, including differentiated cells, rather than a single entity. Stem cells can be isolated from embryos, fetuses or from adult tissue, but the range of cell types to which they can differentiate varies. Embryonic stem cells are the most pluripotent, i.e. they have the potential to become most different types of cells under the right conditions. For tissue engineering, stem cells potentially can provide a virtually inexhaustible cell source.

Current research is focussed on promoting stem cell differentiation to required lineages, purification of consequent cells, confirmation that there is no residual carcinogenic potential in the cell population and implantation in a form that will replace, or augment the function of, diseased or injured tissues. An initial step is the selection of the most appropriate stem cell to form the required tissue.

'Adult' stem cells

Everyone carries around their own repository of stem cells that exist in various tissue niches, including bone marrow, brain, liver and skin, as well as in the circulation. Originally, these cells were considered to have only oligolineage (monopotent) potential but it is now known that they can show a considerable degree of plasticity. In theory, therefore, these cells could be removed from a patient, incorporated into a tissue construct and put back into the same individual when repair becomes necessary, bypassing the need for immunosuppression. Clearly, adult-derived progenitor cells need to be investigated and their clinical usefulness established. However, as mentioned above, for some stem cell types, problems with accessibility (e.g. it is not easy or desirable to tap stem cells in the brain!), low frequency (e.g. in bone marrow there is roughly 1 stem cell per 100 000 cells and this might also be affected by age and disease), restricted differentiation potential and poor growth may limit their applicability to tissue engineering.

Embryonic stem cells

Embryonic stem cells, whilst raising ethical concerns in some quarters, remain the most plastic cell source available to tissue engineers. Murine embryonic stem (ES) cells were first described more than two decades ago, when they were isolated from the inner cell mass of the developing blastocyst (the early development phase of an embryo) and grown in the laboratory. ES cells have since been shown to be totipotent, differentiating to all lineages, including the germline and trophoblast. *In vitro*, murine ES cells were shown to proliferate indefinitely in the undifferentiated state and retain the capacity to differentiate to all mature somatic phenotypes when they received the appropriate signals. The initial isolation of ES cell lines heralded a major breakthrough for developmental biology as it provided a simple model system for studying the processes of early embryonic development and cellular differentiation. However, it also opened the way for tissue engineering applications; if ES cells could be derived from human blastocysts, their capacity for multilineage differentiation might be exploited, e.g. for cell-based therapies in which virtually any tissue or cell type could be produced on demand in the laboratory. Human ES cells were eventually derived in 1998, providing a tremendous boost for tissue engineering.

Human ES cells show several important differences from murine ES cells *in vitro*. Human ES cells tend to grow more slowly, usually forming flat, rather than spherical, colonies and are dissociated more easily into single cells than their murine counterparts. Unlike murine, human ES cells are also unresponsive to leukaemia inhibitory factor (LIF). Therefore, to keep them undifferentiated, they require culture on murine embryonic fibroblast feeder (MEF) feeder layers in the presence of basic fibroblast growth factor or on specific substrates such as matrigel or laminin in MEF-conditioned medium.

Differentiation of ES cells is often initiated via the formation of distinct cellular aggregates or embryoid bodies (EBs), which are made up of cells of the three basic germ layers; ectoderm, endoderm and mesoderm. Most ES cells form EBs spontaneously and they have proved useful in studying early events in mammalian development. Within just a few days, EBs can grow to comprise many thousands of cells and are a rich source of progenitor cells of all germ layers. Progenitor cells (sometimes called transit-amplifying cells) are 'committed' to forming a particular cell type. They retain a finite capacity to proliferate and help define the 'bulk', structure and function of tissues.

18.4   Culture conditions

As indicated above, by manipulating the culture conditions under which stem cells are maintained, it is possible to control or restrict the available differentiation pathways and selectively generate cultures enriched with a particular phenotype. Such manipulations include stimulation of cells with particular cytokines, growth factors, amino acids, other proteins and active ions and co-culture with the target cell/tissue type. Often, cell sorting techniques such as fluorescence-activated cell sorting (FACS) are used to purify further a particular cell type. Utilising these approaches, virtually every cell type in the body has been derived from stem cells, mainly *in vitro* but stable stem cell-derived grafts have been established *in vivo*.

Stem cells are also amenable to genetic manipulation, in particular ES cells, and have been instrumental in the creation of transgenic and gene knockout animals permitting more detailed investigation of the genome and the specific functions of a particular gene. This genetic tractability also offers the potential to introduce genes to promote lineage-restricted differentiation and provides a basis for gene therapy to introduce therapeutic genes and, potentially, to modulate the immune response allowing implantation of 'non-self' cells/tissues.

Chapter 20 describes the cell culture techniques that can be employed in the development of tissue-engineered constructs.

18.5   Three-dimensional interactions

The normal function of most cells and tissues is, in addition to soluble factors, dependent on spatial interaction with neighbouring cells and with a substratum or extracellular matrix (ECM). Cell–cell and cell–ECM interactions are co-ordinated by members of several families of membrane spanning proteins, called adhesion molecules. These are fundamental to cell adhesion, helping to define 3-D cellular organization and also to participate directly in cell signalling, controlling cell recruitment, growth, differentiation, immune recognition and modulation of inflammation. Consequently, recapitulating the function of the ECM and 3-D cell interactions is an important aspect of generating viable constructs for *in vivo* tissue replacement.

A number of natural and synthetic materials have been used to produce 3-D scaffolds to function as an artificial ECM (Chapter 19). Scaffolds for tissue repair ideally should be non-toxic, have good biocompatibility, be biodegradable and be capable of interacting specifically with the cell type(s) of interest. Work with such materials has shown how scaffolds can also be made to be bioactive through adsorption with biomolecules and that such modifications can enable specific recruitment and adhesion of specific cell types.

18.6   Cell reprogramming

In the light of the success of cloning Dolly the sheep, at the Roslin Institute, and the various other animals that have followed, much interest has been generated in understanding the mechanisms of nuclear cloning/reprogramming and potentially harnessing them for tissue repair strategies. In nuclear cloning, an enucleated oocyte is fused with the nucleus of a somatic cell. This stimulates 'rewinding' of the genetic programme of the somatic nucleus to generate a totipotent cell. This cell can be used to generate an ES cell line that could be used to generate specific cell/tissue types that would be genetically identical to the donor somatic cell (therapeutic cloning). If the egg is implanted into a surrogate mother, it can potentially form an intact embryo (reproductive cloning). The factors and mechanisms that induce this remarkable transformation are not known but it seems likely that a component, e.g. cytokine, hormone etc. of the enucleated oocyte stimulates this reprogramming process.

18.7   The way forward

In this chapter, we have discussed some of the most recent developments in the use of stem cells for tissue repair and regeneration. There is no doubt that stem cells derived from adult and embryonic sources hold great therapeutic potential but it is clear that there is still much research required before their use in the clinic is commonplace.

As mentioned above, there is much debate about whether adult stem cells can be used instead of ES cells. The opinion of these authors and many others working in this field is that it is too early to disregard one or other of these cell sources (see Table 18.1, which lists a few 'pros' and 'cons' associated

Table 18.1 Comparisons of ES cell and 'adult' stem cells and application to tissue repair

	ES cells	Bone marrow stromal stem cells
In vitro proliferation	Indefinite	Unclear, probably finite ~ 50 population doublings
Stability	Karyotype may change with prolonged culture	Niche may change with age or disease affecting cell numbers and differentiation
Accessibility	Several established cell lines. Consistency of lines likely but not confirmed	Some cell lines, but generally requires aspiration of tissue. May be variations associated with sample site/technique and accessibility
3-D interactions	Typically grow as aggregates and differentiate as EBs	Generally monolayer cultures
Repair *in vivo*	Yes, a few animal studies. Existing lines can not be used in humans	Yes, shown in many animal studies, autologous and allogeneic transplants in humans

with ES cells and 'adult' stem cells). There is certainly a need for rationalisation but this can only be exercised once we have carefully compared and contrasted these various cells under the appropriate experimental conditions. Some characteristics that might help resolve the issue of cell source can already be applied to the debate. Accessibility of cells is obviously important. In terms of adult stem cells, it is already clear that some cells, like neural stem cells, pose significant difficulties in harvesting (at least in living donors). Even cells that are more accessible, such as marrow stem cells, are harvested using an invasive aspiration procedure. There are also issues concerning the incidence/ abundance of adult stem cells and their numbers and potency that might decline with increasing age or be affected by disease.

With regard to ES cells, concerns have been raised by certain religions, and the proposed practice of therapeutic cloning tends to be misrepresented in the media. The creation of ES cells can be offset somewhat by the fact that there are potentially 'large' numbers of embryos created by *in vitro* fertilisation programmes that are surplus to requirements (normally destroyed) and could potentially be used for derivation of ES cells. In the UK the government has, with considerably foresight, paved the way for extensive but carefully monitored stem cell research through its legislation and the creation by the Medical Research Council (MRC) of a stem cell bank.

As a final word of warning, for both adult and embryonic stem cells, their stability, potential to transmit harmful pathogens or genetic mutations, risk of forming unwanted tissues or even teratocarcinomas have yet to be fully evaluated.

18.8   Summary

In this chapter, some of the key challenges facing tissue engineers are addressed. Cell sources are discussed, including autologous primary and embryonic and adult stem cells, and the means by which the cells can be propagated and encouraged to differentiate towards specific lineages. The requirements for scaffolds to produce 3-D tissue constructs are outlined, and the contentious issue of stem cell cloning is presented in the context of its applications in tissue engineering. Finally, an appraisal of the future of tissue engineering is given, concentrating on the potential applications of stem cells.

18.9   Key definitions

Blastocyst: early development phase of an embryo.

Cytokines: regulatory proteins that are released by cells of the immune system and act as intercellular mediators in the generation of an immune response.

Ectoderm: outermost of the three primary germ layers of an embryo, from which the epidermis, nervous tissue and sensory organs develop.

Endoderm: innermost of the three primary germ layers of an embryo, developing into the gastrointestinal tract, lungs and associated structures.

Karyotype: characterisation of the chromosomal complement of an individual or a species, including number, form and size of the chromosomes.

Mesoderm: middle embryonic germ layer, from which connective tissue, muscle, bone, and the urogenital and circulatory systems develop.

Oocyte: cell from which an ovum develops (a female gametocyte).

Phenotype: characteristics of a specific cell type.

Plasticity (of cells):   the potential for cell differentiation.

Pluripotent: a pluripotent stem cell has the potential to differentiate into several cell types.

Somatic cell: any type of cell that is not involved in reproduction.

Stromal cell: a cell with a structural function, found in bone marrow, that supports haemopoietic (blood-generating) cells.

Teratocarcinoma: a malignant tumour consisting of different types of tissue, caused by the development of independent germ cells.

Totipotent: a totipotent stem cell has the potential to differentiate to any cell type.

Trophoblast: a cell of the outermost layer of the blastocyst (trophoderm).

18.10   Reading list

Atala A. and Lanza R., *Methods of Tissue Engineering*, Philadelphia, Academic Press, 2001.
Hollinger J.O., *Bone Tissue Engineering*, Boca Raton, Florida, CRC Press, 2004.
Lanza R., Langer R. and Vacanti J. (eds), *Principles of Tissue Engineering*, Philadelphia, Academic Press, 2000.
Marshak D.R., Gardner R.L. and Gottlieb D. (eds), *Stem Cell Biology*, New York, Cold Spring Harbor Laboratory Press, 2001.
Palsson B. (ed), *Tissue Engineering*, London, Prentice Hall, 2004.

The following websites provide useful information on the principles and ethics of stem cells:
http://www.roslin.ac.uk
http://www.doh.gov.uk/cegc/index.htm
http://www.royalsoc.ac.uk/policy/
http://www.nih.gov/news/stemcell/primer.htm
http://www.imperial.ac.uk/medicine/is/tissue

19

Scaffolds for tissue engineering

JULIAN R. JONES
Imperial College London, UK

19.1   Introduction

All present-day orthopaedic implants lack three of the most critical characteristics of living tissues: (1) the ability to self-repair; (2) the ability to maintain a blood supply; and (3) the ability to modify their structure and properties in response to environmental factors such as mechanical load. All implants have a limited lifespan and, as life expectancy is continually increasing, it is proposed that a shift in emphasis from *replacement* of tissues to *regeneration* of tissues is required to satisfy this growing need for very long-term orthopaedic repair.

Tissue engineering was introduced in Chapter 18 as the growth of tissue-like construct in the laboratory, ready for implantation to regenerate a diseased or damaged tissue to its original state and function. It is a multi-disciplinary field that involves cell and molecular biology, materials science, chemical and mechanical engineering, chemistry and physics. There is the potential for stem cells to be extracted from a patient, seeded on a scaffold of the desired architecture *in vitro*, where they will be given the biological signals to proliferate and differentiate, and the tissue will grow, ready for implantation. This chapter concentrates on the development of such scaffolds and discusses how the materials used should be tailored for the regeneration of specific tissues. In any scaffold application, the material choice and design is important.

19.2   Classes of potential scaffold materials

All materials elicit a biological response from the body when they are implanted. Many are toxic to the body, but many are also biocompatible (not toxic). There are three classes of biocompatible materials: bioinert, resorbable and bioactive.

Bioinert materials

No material is completely inert on implantation, but the only response to the implantation of bioinert materials is encapsulation of the implant by fibrous tissue (scar tissue). Examples of bioinert materials are medical grade alumina, zirconia (Chapter 3), stainless steels (Chapter 2) and high-density polyethylene (Chapter 4) that are used in the total hip replacements (Chapter 13).

Resorbable materials

Resorbable materials are those that dissolve in contact with body fluids, and the dissolution products can be secreted via the kidneys. The most common biomedical resorbable materials are polymers that degrade by chain scission, such as polyglycolic (PGA) and polylactic acids (PLLA) and their co-polymers, which are commonly used as sutures (Chapter 10). Some bioceramics are also resorbable *in vivo*, such as calcium phosphates (Chapter 3).

Bioactive materials

Bioactive materials stimulate a biological response from the body such as bonding to tissue. There are two classes of bioactive materials: osteoconductive and osteoproductive. Osteoconductive materials bond to hard tissue (bone) and stimulate bone growth along the surface of the bioactive material, e.g. synthetic hydroxyapatite and tri-calcium phosphate ceramics. Osteoproductive materials stimulate the growth of new bone on the material away from the bone/implant interface, e.g. bioactive glasses, which can also bond to soft tissue such as gingival (gum) and cartilage.

   The mechanism of bone bonding to bioactive materials is thought to be due to the formation of a hydroxyapatite layer (HA) on the surface of the materials after immersion in body fluid. This layer is similar to the apatite layer in bone and therefore a strong bond can form. The layer forms quickest on osteoproductive materials.

19.3   The criteria for an ideal scaffold

There are many criteria that must be fulfilled to create an ideal scaffold. Such a scaffold has yet to be developed, but it may not be far away.

   To be able to regenerate a tissue, the scaffold should have a structure that acts as a template for tissue growth in three dimensions and stimulates new growth in the shape dictated by the scaffold. The obvious design for a template is a structure that mimics the structure of the host tissue. To allow a tissue to grow in 3-D, the template must be a network of large pores (macropores).

The pores must be connected to each other and the apertures between the pores must have diameters in excess of 100 μm. The interconnected pore network is necessary to allow cells to migrate through the scaffold and promote tissue growth throughout the template. During *in vitro* cell culture and tissue growth, the pore network will allow the culture media to reach all cells, providing them with essential nutrients. Once the tissue-engineered construct has been implanted, the pores must be connected sufficiently such that blood can penetrate to provide those nutrients. Eventually, an ideal scaffold would stimulate blood vessels to grow inside the pore network (angiogenesis). The minimum aperture diameter for angiogenesis, and for bone ingrowth for bone tissue engineering applications, is 100 μm.

It is not only the macropore morphology that needs to be optimised but also the surface topography. Many cells, such as osteogenic cells, must attach to a substrate before they lay down their extracellular matrix. A surface texture of nanometre-sized pores or surface roughness on the nanometre scale can increase cell activity.

When bioinert materials are implanted, a fibrous tissue (scar tissue) surrounds the implant. If a bioinert scaffold is implanted, it would be isolated rapidly from the host tissue and the host tissue will not regenerate. The scaffold material should therefore, be able to bond to the host tissue without the formation of scar tissue. One way to achieve bioactivity is to use a bioactive material such as a bioactive ceramic or glass. For bone tissue engineering applications, an osteoconductive material such as HA may be suitable. If higher bioactivity or osteoproduction is required, then bioactive glasses can be used. Certain compositions of bioactive glass can also bond to soft tissue. Polymers that are not bioactive can be made to bond to soft tissue by modifiying their surface and attaching proteins such as laminin to the surface of the polymer, which will encourage tissue adhesion.

The mandate for tissue engineering is to regenerate a tissue to its original state and function. This cannot happen if there is still artificial material left in the host tissue. An ideal scaffold should therefore be resorbable so that eventually there is no trace of the scaffold's presence. The dissolution products of the scaffold should be ones that can be metabolised by the body. The degradation rate of the scaffold must be controllable so that it can be tailored to match the rate of tissue growth once it has been implanted. Resorbable polymers and hydrogels are the most commonly used resorbable scaffolds, but tricalcium phosphate and certain bioactive glasses offer both resorbability and bioactivity.

An ideal scaffold would not only act as a template for tissue growth and have controlled resorbability, but it should also activate the cells of the tissue for self-regeneration. The scaffold should act as a delivery system for the controlled release of cell- and gene-stimulating agents. The agents can be a number of different substances. Growth factors can be incorporated into resorbable polymer (Chapter 10) or hydrogel (Chapter 11) scaffolds, which will release the growth factors into the body as the scaffold resorbs. Alternatively, bioactive glasses and silicon-substituted HA scaffolds release small concentrations of silicon and calcium ions, which have been found to stimulate seven families of genes in osteoblasts, increasing proliferation and bone extracellular matrix production.

Importantly, the mechanical properties of the tissue-engineered construct should match that of the host tissue. This is evident from the history of total hip replacements (Chapter 13). The metal alloy implants have a much higher stiffness (Young's modulus) than the bone. This causes an effect called stress shielding, where all load is transferred though the metal and not the bone, causing the body to resorb the bone. This is one of the main disadvantages of using bioactive ceramics and glasses for tissue regeneration. However, if tissue is grown on a bioceramic scaffold *in vitro*, the tissue-engineered construct is then a biocomposite and its modulus will decrease. Only the mechanical properties of the final tissue-engineered construct that will be implanted are critical. Completely artificial composites (Chapter 5) and hybrid materials are being developed to optimise the mechanical properties of scaffolds. The structure and strength of a resorbable scaffold must also be maintained until enough host tissue has been regenerated.

If the scaffold is to be mass produced, so that surgeons can use it in the clinic, it must be made from a processing technique that can be up-scaled for mass production and that can produce an irregular shape to match that of the defect in the bone of the patient (Chapter 24). The scaffold must also have the potential to be producible to the required International Standards Organisation (ISO) or Food and Drug Administration (FDA) standards and be easily sterilised (Chapter 23).

The need for implant materials to pass FDA may be necessary for the safety of patients, but it places a large financial obstacle in the path of new scaffold developments. Many researchers use materials that have already been passed by the FDA so that they are sure their scaffold will be safe to use in the clinic. They are unwilling to risk using materials that have not undergone FDA testing because of the time and financial commitment required to fulfil the extensive tests.

19.4   Polymer scaffolds

Resorbable polymers are a popular choice of material for tissue-engineered scaffolds for three reasons. Firstly, polymers are easy to process in the shape of a 3-D scaffold with a pore morphology suitable for tissue engineering applications. Secondly, polymers can have high tensile properties and high toughness and the mechanical properties of polymers can be controlled very easily by changing the molecular weight (chain length) of the polymer.

Thirdly, bioresorbable polymers have been used successfully as dissolving sutures for many years. Therefore, these degradable polymers, such as the polyesters of poly(lactic acid) (PLA), poly(glycolic acid) (PGA) and poly(lactic acid-co-glycolic acid) (PLGA) are used for scaffold applications because they have passed FDA regulations, and scaffolds made from these materials can provide a quick route to a commercial and clinical product (Chapter 10).

The techniques used to produce porous networks in these polymers are fibre bonding or weaving, solvent casting, particulate salt leaching, phase separation, gas foaming, freeze drying and extrusion.

For many techniques, the polymer must be dissolved in a solvent. One technique using the polymer solution is salt leaching, where salt particles are dispersed throughout the solution before the polymer is cast. The salt particles can then be dissolved out of the matrix-leaving pores but this technique only works for thin membranes or for 3-D specimens with very thin wall sections: otherwise, it is not possible to remove the soluble particles from within the polymer matrix.

The polymer solution can also be foamed to produce an open pore structure. This can be achieved by using blowing agents, gas injection, supercritical fluid gassing or freeze–drying.

The polymers that can be used for supercritical fluid gassing must have an high amorphous fraction (Chapter 4). Polymer granules are plasticised due to the use of a gas, such as nitrogen or carbon dioxide, at high pressures. The dissolution of the gas into the polymer matrix results in a reduction of the viscosity, which allows the processing of the amorphous bioresorbable polyesters in a temperature range of 30–40 °C. However, on average, only 10–30% of the pores are interconnected.

During freeze–drying, the polymer is dissolved in a solvent and frozen in liquid nitrogen while the growth direction of ice is controlled. The frozen solution is then dried under reduced pressure to cause sublimation of the ice, creating pores macroscopically aligned in the direction of freezing. The green bodies are then sintered to provide strength. Large orientated pores with porosities in excess of 90% can be obtained. Figure 19.1 shows an SEM micrograph of freeze–dried PGLA foam scaffold.

There are, however, some problems with using biodegradable polymers as scaffolds for bone regeneration. A degradable scaffold is desired; however, sutures dissolve within 2 weeks of implantation, which is too rapid for bone regeneration applications. Secondly, although resorbable polymers can be made with high tensile strength and toughness and their mechanical properties can be matched with collagen, their Young's modulus is much lower that that of bone. Therefore, these polymers cannot be used in load-bearing sites where they undergo compressive forces.

A third problem is that the mechanical strength of polymers decreases rapidly as they degrade. High tensile strength polymers have long chains

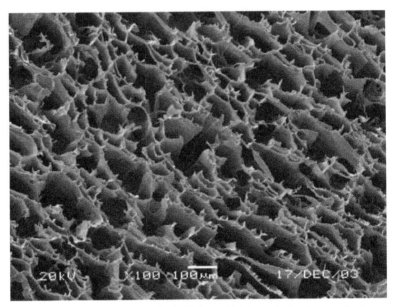

19.1 Electron micrograph of a PGLA scaffold made by freeze–drying.

(high molecular weight), which entangle with each other (Chapter 4). As a tensile force is applied to the entangled chains, the chains begin to unravel until they become straight. The maximum tensile strength of polymers is reached when the chains are fully extended. Therefore, for a polymer to be tough, it must have a molecular weight over the entanglement value. The mechanism for degradation of biodegradable polymers is chain scission. The polymers undergo hydrolysis and the chains are cut in two. The average molecular weight of the polymer is therefore approximately halved with every chain scission event. The mechanical properties of the polymer are proportionate to the square root of the molecular weight and therefore decrease very rapidly as the polymer degrades. Biodegradable polymers have also been found to produce an inflammatory response due to the acidic byproducts of the degradation.

Generally, polymers are suited for the engineering of soft tissues such as the skin and connective tissues such as ligaments. Natural polymers, such as collagen, gelatin and fibrin, can also be used as tissue-engineered scaffolds. Recombinant human collagens are an efficient scaffold for bone repair when combined with a recombinant bone morphogenic protein in a porous, sponge-like format. They can also be used as a membrane, sponge or hydrogel and can serve as a basis for the engineering of skin, cartilage and periodontal ligaments, depending on the specific requirements of the chosen application.

Highly porous copolymers of type I collagen and chondroitin 6-sulphate (a glycosaminoglycan or GAG) have been produced. They are used as dermis regeneration scaffolds, where they are used as a graft to cover deep and extensive skin wounds. They are also used for nerve regeneration scaffolds where they are used as filling for a nerve chamber in which the two nerve stumps of a transected nerve are inserted.

Hydrogels can be applied to tissue-engineering applications in ways different from other polymers as cells can grow within the gel matrix. Pore networks are therefore not always required, but pores can be introduced by the freeze–drying process. Hydrogels and their tissue engineering applications are discussed in Chapter 11.

19.5   Bioactive ceramic scaffolds

Ceramics are crystalline materials (Chapter 3) and therefore tend to have high compressive strength and Young's modulus but low toughness, i.e. they are brittle materials. Alumina and synthetic HA are the ceramics that are most commonly used in biomedical applications. Alumina is a bioinert ceramic, so it is not ideal for use as a regenerative scaffold.

HA and TCP have been used successfully in the clinic as bone filler materials in powder form. Synthetic HA has been used most regularly because it has a similar composition, structure and Young's modulus to bone mineral. Tricalcium phosphate (β-TCP) is also similar to bone mineral in that they are both calcium phosphate ceramics, but β-TCP is resorbable. The challenge is to develop them into 3-D porous scaffolds, i.e. into a open pore structure, with the properties listed in Section 19.3.

In a porous form, HA and β-TCP ceramics can be colonised by bone tissue. A problem with introducing pores into a ceramic is that the compressive strength of the material decreases dramatically. The strength of the scaffold depends on the thickness and strength of the struts or pore walls. Generally for brittle compression of a foam,

$$\sigma_{cr} \propto \rho_r^{3/2}$$
[19.1]

where σ_{cr} is the critical strength of the pore walls and ρ_r is the relative density, where

$$\rho_r = \rho_b/\rho_s$$
[19.2]

where ρ_b is the bulk density of the scaffold and ρ_s is the skeletal (true) density of the material.

The simplest way to generate porous scaffolds from ceramics such as HA is to sinter particles, preferably spheres of equal size. The scaffolds can then be pressed using cold isostatic pressing. As sintering temperature increases, pore diameter decreases and mechanical properties increase as the packing of the spheres increases. Mechanical properties can be increased further by hot isostatic pressing (HIPing). Porosity can be increased by adding fillers such as paraffin, naphthalene, hydrogen peroxide, sucrose, gelatine and polymethyl methacrylate (PMMA) microbeads to the powder slurry, which burn out on sintering, leaving pores. However, this technique generally decreases the compressive strength to below that of trabecular bone.

The majority of methods that are used to produce polymer foams cannot be applied to ceramic systems. However, a popular method for producing highly porous ceramics is to produce a polyurethane foam template that can be immersed in ceramic slurries under vacuum to allow the slurry to penetrate into the pores of the foam. The organic components are burnt out and the ceramic foams sintered (1350 °C), producing a scaffold with interconnected pore diameters of up to 300 μm.

Ceramic slurries can also be foamed to obtain pores in the range of 20 μm up to 1–2 mm. The incorporation of bubbles is achieved by injection of gases though the fluid medium, mechanical agitation, blowing agents, evaporation of compounds or by evolution of gas by an *in situ* chemical reaction. A surfactant is generally used to stabilise bubbles formed in the liquid phase by reducing the surface tension of the gas–liquid interface. Surfactants are macromolecules composed of two parts, one hydrophobic and one hydrophilic. Owing to this configuration, surfactants tend to adsorb onto gas–liquid interfaces with the hydrophobic part being expelled from the solvent and a hydrophilic part remaining in contact with the liquid. This behaviour lowers the surface tension of the gas–liquid interfaces, making thermodynamically stable the foam films, which would otherwise collapse in the absence of surfactant.

The gel-casting method has been the most successful method used to produce macroporous bioactive hydroxyapatite (HA) ceramics with interconnected pores of greater than 100 μm in diameter. Suspensions of HA particles and organic monomers are foamed by agitation with the addition of a surfactant under a nitrogen atmosphere. *In situ* polymerisation (cross-linking) of the monomers, which is initiated by a catalyst, creates a 3-D polymeric network (gel). The porous gels are sintered to provide mechanical strength and to burn out the organic solvents. HA foams have been made with compressive strengths in excess of 10 MPa, which is similar to that of trabecular bone. When the foams were implanted into the tibia of rabbits, bone partially filled the pores after 8 weeks and there was no inflammatory response. The compressive strength trebled in the scaffolds that were colonised by bone.

Porous HA substrates produced by the above methods are available commercially, such as Endobon® (Merck, Darmstadt, Germany) and Interpore® (Interpore International, Irvine, CA, USA).

Naturally occurring porous structures are being considered for fabricating HA scaffolds. A frequently used structure is coral. Hydrothermal and solvothermal methods are used to transform natural coral into HA after removal of the organic component by, for example, immersion in sodium hypochlorite. Pore size in typical coral formations is in the range 200 to 300 μm. The porosity is interconnected and the structure resembles that of trabecular bone.

An HA scaffold does not satisfy the criteria of an ideal scaffold because HA resorbs only very slowly, the dissolution products do not stimulate the genes in the osteogenic cells and HA is only osteoconductive, it is not osteoproductive and does not bond to soft tissue. HA is still a bone replacement material rather than a regenerative material. It is possible, however, to modify the HA composition to achieve gene activation. Genes are activated by small concentrations (less than 20 parts per million) of hydrated silicon ($Si(OH)_4$), therefore, small amounts of silica (SiO_2) can be substituted for calcia (CaO) in synthetic hydroxyapatite. *In vivo* experiments have shown that bone ingrowth in silica-substituted HA granules was significantly greater than that into phase-pure HA granules.

Silicon-substituted HA scaffolds therefore have the potential to fulfil all criteria of an ideal scaffold for bone tissue engineering apart from controlled resorbability and a Young's modulus match with the host bone. They are generally unsuitable for soft tissue engineering.

19.6   Bioactive glass scaffolds

Bioactive glasses form a bond to bone much quicker than bioceramic materials. This is due to them being amorphous. Dissolution of a random amorphous network begins much earlier than that of a crystalline ceramic; therefore, the HA forms more quickly on the glass than on synthetic HA. Bioactive glasses can be made via two routes: traditional melt-processing and the sol–gel process. Sol–gel-derived bioactive glasses have a porous texture in the nanometre range, giving them a surface area of 150–600 m^2g^{-1}, which is two orders of magnitude higher than melt-derived glasses. Dissolution is therefore, more rapid than for melt-derived bioactive glasses of similar composition. Along this surface there are many silanol groups that act as nucleation sites for HA layer formation, also making sol–gel-derived glasses more bioactive. The dissolution of sol–gel-derived bioactive glasses is large enough for bioactive glasses to be considered resorbable and the rate of resorption can be controlled by altering the textural porosity.

Some of the techniques used to produce porous ceramics can be applied to bioactive glasses but the results are variable. Sacrificial porogens and foaming agents have also been used to create melt-derived bioactive glass (Bioglass®) scaffolds. Large pores with diameters in the region of 200–300 μm were created but the total porosity was just 21% and there were large distances between pores, so this process failed to mimic the interconnectivity of trabecular bone.

Theoretically, gel-casting could be applied to melt-derived bioactive glass

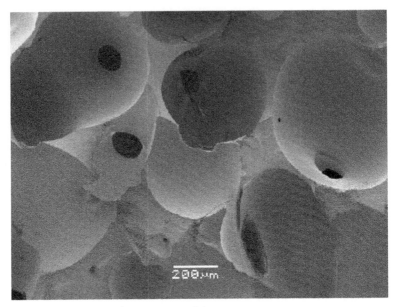

(a)

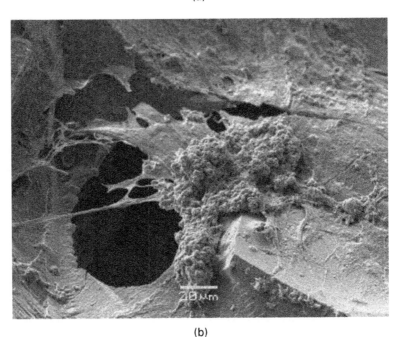

(b)

19.2 SEM micrographs of (a) a bioactive glass foam scaffold and (b) primary human osteoblasts, cultured on a bioactive glass foam scaffold for 2 weeks, which have produced mineralised bone nodules.

powders. However, bioactive glasses undergo surface reactions on contact with solutions to produce an HCA surface layer and it is desirable to have control over the reaction before a scaffold is ready for clinical use.

The most successful method for the production of bioactive glass scaffolds of similar structure to trabecular bone mineral is the foaming of sol–gel-derived bioactive glasses. The sol–gel process involves polymer reactions of the glass precursors in a solution (sol). The sol is a solution of silica species that forms a gel by cross-linking together into a silica network.

During the foaming process, air is entrapped in the sol under vigorous agitation as viscosity increases and the silica (–Si–O–Si–) network forms. A surfactant is added to stabilise the bubbles at short times. As the porous foam becomes a gel, the bubbles are permanently stabilised. The gel is then subjected to controlled thermal processes of ageing (60 °C) to strengthen the gel, drying (130 °C) and thermal stabilisation/sintering to remove organic species from the surface of the material (500–800 °C). Bioactive glass foam scaffolds can contain macropores up to 600 μm in diameter, connected by pore windows with modal diameters in excess of 100 μm and compressive strengths up to 2.5 MPa. Figure 19.2(a) shows a scanning electron micrograph of a pore network of a typical bioactive glass foam. *In vitro* cell studies using primary human osteoblasts have shown that the foams stimulate formation and mineralisation of bone nodules within 2 weeks of culture (Fig. 19.2(b)). The only criterion of an ideal scaffold that is not fulfilled by bioactive glass foam scaffolds is the toughness in tension.

19.7   Composites

Bone is a natural composite of collagen (polymer) and bone mineral (ceramic). Collagen fibres have a high tensile and flexural strength and provide a framework for the bone. Bone mineral is an apatite that provides stiffness and the high compressive strength of the bone. The obvious method to mimic the structure of bone would be to create a porous scaffold from a composite material (Chapter 5).

Composites are generally a polymer matrix with ceramic or glass fibres or particles reinforcing the matrix. Attempts to produce an ideal scaffold from composite materials has led to the use of the PGA/PLLA and poly(caprolactone) resorbable polymers as described in Section 19.4. Filler materials of bioactive ceramics such as synthetic HA and tricalcium phosphate have been added to impart not only increased strength and stiffness but also bioactivity to the polymer matrix. Bioactive glass particles have been incorporated into PLGA foam scaffolds shown in Fig. 19.1 to provide not only bioactivity but also delivery of gene activating ions to the scaffold (CD Fig. 19.3). It is still very difficult to maintain the resorbability and maintain the mechanical properties of the scaffold during resorbtion, so the quest for an ideal scaffold continues.

19.8   Control of architecture

The above sections have discussed material selection and techniques to produce an interconnected porous network. The foaming techniques produce suitable pore networks, but although these techniques allow control of the percentage porosity and mean pore diameter, the individual pore architectures and position cannot be controlled. Some tissue engineering applications need a graded porosity, e.g. a multiple tissue interface such as an articular cartilage/bone transplant.

Ideally, a scaffold should be designed for a specific defect site in a patient. An X-ray computer tomography (CT) scan of a damaged tissue can provide the exact architecture and shape of a scaffold required to aid tissue regeneration. This data can be fed into computer-aided design (CAD) files. These files can then be used to dictate the pore network structure of the scaffold by using solid free-form fabrication (SFF) and rapid prototyping (RP) techniques. SFF and RP refer to a variety of technologies capable of producing 3D physical structures from 3D computer data sets. The use of SFF for scaffold fabrication, mainly for polymeric scaffolds, has steadily increased over the past few years. Among existing systems, stereolithography, extrusion free-forming or fuse deposition modelling, inkjet printing and selective laser sintering methods are possible choices. Figure 19.3 (CD Fig. 19.5) shows a polymer scaffold that has been created by fused deposition modelling. This technique uses a filament of polymer that is heated and laid on a substrate in

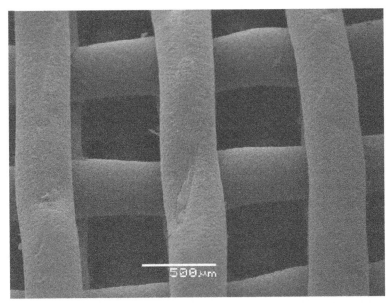

19.3 A polymer scaffold that has been created by fused deposition modelling (CD Fig. 19.5).

a layer-by-layer process through a nozzle, the movements of which are dictated by a CAD file.

RP of bioceramics is used in two different ways: the direct route in which the scaffold is directly formed from the CAD file to the final form, and the indirect route in which first a negative of the desired structure is made with polymer wax, which acts as a mould. The moulds are filled with a HA aqueous slurry, prepared conventionally, using a vacuum infiltration device. Pyrolysis of the mould and final sintering of the ceramic scaffolds are then carried out at 1250 °C. A perspective view of a typical scaffold fabricated by this technique is shown in Fig. 19.3. The typical flow diagram for HA scaffold design and manufacturing by indirect RP techniques is schematically shown in CD Fig 19.5, which involves three main steps: (a) mould microstructure design, (b) ceramic slurry development, and (c) binder burn out and sintering.

Although less common, direct RP of HA scaffolds are possible using inkjet printing systems and by the commercially known TheriForm™ SFF process. In this process a thin layer of powder is spread onto a build bed, while a computer-controlled print-head assembly deposits liquid binder droplets onto selected regions of the powder. The floor of the powder bed drops down, a new layer of powder is spread, and the process is repeated until construction of the object is complete.

Extrusion free-forming is another promising technique used to produce hydroxyapatite scaffolds with highly controlled pore structure (CD Fig. 19.6). Bioactive glasses have not yet been used in RP or SFF techniques.

19.9   Summary

Numerous materials have been developed as scaffolds for tissue engineering applications. Scaffolds that most closely match the criteria for an ideal scaffold and most closely mimic the structure of trabecular bone are made by foaming. These include gel-cast HA foams, bioactive glass foam and biodegradable polymer foam/ bioactive glass composites. However, not all the criteria of an ideal scaffold have been achieved by any material or processing technique. Scaffolds produced by RP and SFF methods exhibit highly ordered microstructures and they can be manufactured readily to complex shapes dictated by CT scans direct from the patient. A combination of polymers and bioceramics with newly generated tissue must be used to attain the goal of an ideal scaffold.

19.10   Reading list

Banner N., Polak J.M. and Yacoub M., *Lung Transplantation*, Cambridge University Press, 2002.

Brinker C.J. and Scherer G.W., *Sol–Gel Science: The Physics and Chemistry of the Sol–Gel Process*, Boston, Academic Press, 1990.

Chalmers J. and Griffiths P.R., *Handbook of Vibrational Spectroscopy, Volume 5: Applications in Life, Pharmaceutical and Natural Sciences*, New York, Wiley, 2001.

Davies J.E., *Bone Engineering*, Toronto, EM^2 Inc., 2000.

Gibson L.J. and Ashby M.F., *Cellular Solids Structure and Properties*, Oxford, Pergamon Press, 1988.

Hastings G.W. and Williams D.F., *Mechanical Properties of Biomaterials*, Chichester, Wiley, 1980.

Hench L.L. and Wilson J., *An Introduction to Bioceramics*, Singapore, World Scientific, 1993.

Park J. and Lakes R.S., *Biomaterials: An Introduction*, 2nd edition, New York, Plenum, 1992.

Wise D.L., *The Biomaterials and Bioengineering Handbook*, New York, Marcel Dekker, 2000.

20

A guide to basic cell culture and applications in biomaterials and tissue engineering

JAMUNA SELVAKUMARAN and
GAVIN JELL
Imperial College London, UK

20.1   Introduction

There are many advantages of using cell culture in the field of biomaterials. It is easy to manipulate and monitor the environment so as to investigate the responses of cells to biomaterials and biomaterial–cell interactions. It also reduces the need for animal testing in various stages of biomaterials development.

There are two main types of cell culture: one is primary culture and the other is continuous culture (cell lines). Primary cultures are obtained directly from animal or human tissues and are cultured either as small pieces of tissues or single cells following isolation from the tissue by enzyme digestion, e.g. trypsin and collagenase. The main disadvantage of primary cultures is that they become senescent (lose their ability to multiply) and may lose some phenotypic characteristics with time. The main advantage of primary cultures is that they retain many of their original characteristics in their limited life span. Continuous cell lines can be maintained in culture either for a limited number of cell divisions or indefinitely. Many of these cell lines are derived from cancerous tissues of patients, while some of these cell lines are transformed into immortal cells using viral oncogenes. These continuous cell lines have the advantage of unlimited availability but have the disadvantage of preserving few of the original cellular characteristics.

Cells derived from solid tissues must attach to a substrate in order to grow, whilst cells derived from blood grow in suspension. Cells in suspension have a round morphology, and cells attached to a substrate show different morphologies depending on their tissue of origin (Fig. 20.1, CD Fig. 20.1).

Once the cells attach to the substrate, they start to divide and multiply to form a complete layer (commonly known as a confluent layer) covering the substrate. Most of the cells, especially primary cells, become contact inhibited and stop growing when they form a confluent layer, but tumour cells tend to form multiple layers. Cells in culture should ideally be sub-cultured (trypsinised or passaged) or seeded onto materials when they are about 70–80% confluent.

215

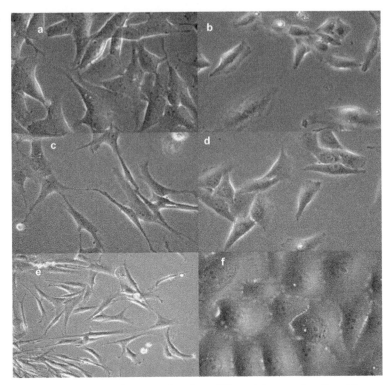

20.1 Phase-contrast images showing different types of cells with distinctive morphologies. (a) SH-SY5Y, human neuroblastoma cell line, (b) A549, human lung carcinoma cell line, (c) primary human fetal osteoblasts, (d) MG-63, human osteosarcoma cell line, (e) primary human fibroblasts, (f) primary human keratinocytes. (a)–(e) magnification 200× and (f) magnification 400×.

Cells require carbohydrates, salts, amino acids, vitamins, fatty acids and proteins to survive *in vitro*. The basal medium contains the essential inorganic salts, amino acids, glucose, vitamins, fatty acids and some proteins. It also contains phenol red and a bicarbonate-based buffering system. Most of the cells in culture require an optimum pH between 7.2 and 7.4. The bicarbonate-based buffering system in combination with a 5% CO_2 atmosphere maintains the optimum medium pH. The phenol red in the medium acts as a pH indicator: it is pink at optimum pH, turns yellow when the medium becomes acidic and purple when the medium turns basic. The basal medium is normally supplemented with 10–15% fetal bovine serum (FBS), which is enriched with proteins and growth factors. A combination of different antibiotics and/ or anti-fungal agents can also be used in cell culture to avoid contamination. The cells are grown in a humid incubator at 37 °C (optimum temperature) with a 5% CO_2 atmosphere. It is necessary to follow sterile procedures when handling cells in culture to avoid microbial contamination.

20.2   Sterilisation

Sterilisation is the inactivation (killing) of living micro-organisms (fungi, protozoa, bacteria, mycoplasma and viruses) from the cell culture environment. Micro-organisms exist everywhere, in the air, on the human body, and on the surface of objects. Successful micro-organism exclusion is therefore essential for meaningful cell culture experimentation. The presence of contamination can cause destruction of cells, reduction of cell growth, alteration of cellular protein expression and contamination of other cell cultures, all of which can lead to grossly misleading results.

Good laboratory practice and basic aseptic laboratory techniques

Good laboratory practice (GLP) should be followed at all times in the laboratory. Guidelines of GLP are found in the Organisation for Economic Co-operation and Development (OECD) for European Union countries. Adhering to the principles of GLP in laboratories ensure that the results are accurate, reproducible and traceable.

GLP guidelines refer to general procedures that allow the successful completion of an experiment in the absence of the principal investigator and ensure experiment repeatability. GLP includes the correct planning of experiments, written records of all experimental procedures completed (in a laboratory book), the training of at least two scientists independently capable of completing the experiment and the proper labelling of all substances/media/culture (name, date, description, passage number and cell type). Written methods or standard operating procedures (SOPs) should always be followed and equipment should be cleaned and properly maintained. GLP also minimises any potential biohazard risk to the investigator.

The general aseptic techniques listed below should always be followed for cell culture:

- Disposable gloves (e.g. nitrile) must always be worn in the cell culture laboratory.
- Clean laboratory coats must always be worn and must not leave the laboratory.
- Work areas and gloved hands must be wiped with 70% alcohol prior to working.
- Work must only be carried out in designated 'clean areas' such as a laminar flow tissue culture cabinet.
- All items and gloved hands must be wiped with 70% alcohol every time the sterile area is entered, e.g. flow cabinet or incubator.
- Sterile containers/flasks must be uncovered only in the flow cabinet immediately prior to use and never left open.

- Sterile disposable pipettes should only be removed from wrappers inside the flow cabinet and only immediately prior to use.
- When caps are removed from sterile containers/flasks, if possible the cap should not be placed on the workbench. The open container should be tilted so that falling micro-organisms fall onto the lip of the bottle.
- Fluids must not be taken from a different bottle with the same pipette; a new sterile pipette must be used for each bottle.
- When performing cell culture procedures there should be no talking, coughing or sneezing.
- Techniques should be performed as quickly as possible to limit contamination.

There are some laboratory procedures where the rules listed above are not possible. It is therefore necessary to be aware that micro-organisms are everywhere and to take steps that will minimise contamination.

Sterilisation methods

There are several methods for sterilising laboratory equipment and/or samples, such as heat, 0.2 μm filtration, chemical cleaning (detergent or alcohol) and radiation (UV light). Most micro-organisms are successfully inactivated (killed) at temperatures of 55–80 °C; therefore, sterilisation techniques that raise temperatures above this value, such as boiling, are commonly used. However, some bacterial endospores and prions can survive these temperatures and therefore, in most cell culture laboratories, autoclaving is the most common and effective technique. Autoclaves produce steam at approximately 134 °C under pressure. Most modern autoclaves also contain a drying function at the end of cycle. The advantages and disadvantages of common sterilisation techniques are given in Table 20.1.

Sterilisation of material samples

Tissue engineering requires cells to grow on a variety of materials or scaffolds and it is essential that these materials are properly sterilised before use. In addition to inactivating micro-organisms, it is also vital that the sterilisation technique does not alter the material properties. For example, many tissue scaffolds are often resorbable and therefore wet sterilisation techniques should not be used, whilst UV light can only sterilise the surface and cannot penetrate through the porous structure. Similarly, heating or chemical sterilisation may change the chemistry, composition and/or mechanical properties of a material, e.g. polycaprolactone, a bioresorbable polymer melts at 60 °C. An additional consideration is the removal of toxic substances used in the sterilisation process, such as alcohol or ethylene oxide gas (ETO).

Table 20.1 The advantages and disadvantages of common sterilisation techniques

Sterilisation technique	Description and applications	Advantages	Disadvantages
Autoclaving	Steam produced at 120–134 °C under pressure. Routinely used for laboratory equipment, i.e. pipette tips, tweezers, etc.	Effective in killing most micro-organisms (with the possible exception of some prions)	Unsuitable for sterilising media, as nutrients are destroyed. May alter material properties. Requires regular pressure checks
0.2 µm filters	Media or other heat sensitive fluids are passed through a sterile 0.2 µm filter	Simple and non-invasive for removal of bacteria and fungi from fluids	Mycoplasma and viruses are not removed. Must be performed in a sterile environment
Ultraviolet light	Normally used to sterilise surfaces and air after cleaning	Inactivates microrganisms and viruses	Poor penetrating power. Not suitable for media sterilisation
Chemical: 70% alcohol	Commonly used as a cleaning agent. Longer exposure time occasionally used in material sterilisation	Low toxicity (compared with other chemical disinfectants), convenient	Not always effective against some fungi, bacterial endospores and some viruses
Chemical: ethylene oxide gas (ETO)	Frequently used for medical materials	Non-damaging to most materials, high levels of sterility	Cannot sterilise liquids, sealed containers or certain plastics. Lengthy (3–7 days), requires specialised equipment
Hot-air ovens (dry heat) (Typically 170 °C for 2 hours)	Routinely used with glassware and metal objects	Items do not get wet	Longer treatments at higher temperature are necessary when compared to autoclaving. Unsuitable for plasticware

In addition to the removal of living micro-organisms, the removal of fragments of dead micro-organisms (in particular bacteria) is also important in material research. Endotoxins are bacterial debris that adhere strongly to materials. These endotoxins can elicit a cellular response in a similar manner to that which would occur in the *in vivo* environment (i.e. if a cell encountered bacteria in the body). When challenged by bacteria, cells typically produce inflammatory mediators to attract immune cells and fight the infection. Therefore, in biomaterial research the washing of samples in addition to sterilisation is often performed.

The importance of proper washing and sterilisation in biomaterial research is highlighted in orthopaedic wear particle experiments. Wear particles are commonly produced from articulating surfaces in orthopaedic implants (hip or knee replacements, Chapter 13). Wear particles are believed to cause an inflammatory response, which eventually results in bone erosion and implant failure. To understand this inflammatory process and to design materials that limit inflammation, the response of cells to wear particles of various chemical compositions, sizes and shapes are studied *in vitro*. However, conflicting reports from different laboratories working on the same area have caused considerable confusion. Recently, several researchers have suggested that the discordant results may be due to the undocumented presence of endotoxins on the wear particles, depending on the relative success or failure of the various sterilisation techniques used. They suggest that the cells are not responding to the particles but rather to the presence or absence of endotoxin. The washing of wear particles in 70% ethanol for 24 hours was shown to successfully remove endotoxins to negligible levels. Endotoxin tests such as the limulus amoebocyte lysate (LAL) assay are commercially available and routinely used in material research.

Testing for cell health and contamination

The procedures listed below should be followed when testing for cell health and contamination:

- Check the culture media. If cloudy a bacterial infection is likely (for adherent cells).
- Determine the number of attached cells in adherent cultures compared with the number of floating cells. Floating cells are generally dead cells.
- Examine cell morphology. Generally (depending on cell type), adherent healthy cells are flattened with extended filopodia (cell extensions), rounded cells are either cells undergoing cell division or are dying cells.
- Look for giant cells (Chapter 7). In most cultures the frequency of giant cells should be relatively low and constant under uniform culture conditions.
- Determine the rate at which the cells in newly established cultures attach

and multiply following passage. Attachment within 1–4 hours suggests that the cells have not been traumatised.

- Whilst fungal and bacterial infections are relatively easy to detect by trained scientists, mycoplasma contamination is not visible and therefore harder to detect. Mycoplasma detection kits are commercially available and should be used if contamination is suspected.

20.3   Cell culture protocols

Most commercially available cells are purchased in frozen form. Upon receipt, these frozen cells must be immediately transferred to liquid nitrogen for storage. The protocol below describes how to resuscitate frozen cells stored under liquid nitrogen.

Resuscitation of frozen cells

1. Prepare growth media.
2. Pre-warm media.
3. Label flask and 15 ml centrifuge tube.
4. Add 8 ml of media to the centrifuge tube.
5. Take the ampoule containing cells out from the liquid nitrogen bank (wear thermal gloves and mask).
6. Leave the ampoule at room temperature for approximately 1 minute (not more than 1 minute).
7. Transfer the ampoule to the water bath at 37 °C and allow it to thaw for approximately 4 minutes.
8. Wipe the ampoule with a tissue soaked in 70% alcohol and transfer immediately to the safety cabinet.
9. Carefully pipette the cells into the centrifuge tube.
10. Rinse the ampoule with 1 ml of fresh media and add to the centrifuge tube.
11. Centrifuge the tube at $200 \times g$ for 5 minutes.
12. Take the supernatant out without disturbing the pellet.
13. Flick the tube to disturb the pellet.
14. Add 1 ml of fresh growth media and make an even suspension of cells by repeated pipetting.
15. Add required amount of growth media to the flask.
16. Add the cell suspension to the flask and mix well.
17. Transfer the flask to the incubator.

When the cells reach sub-confluence levels (70–80% confluent), they must be sub-cultured and seeded onto materials or into flasks at a lower density. This procedure is also known as passaging or trypsinisation. Every time the cells are sub-cultured, they increase in passage number. For experiment reproducibility, the cell passage number should be similar in different cell culture experiments. Increased passage may result in phenotypic changes in cells.

Sub-culture of adherent cells

1. Prepare growth media.
2. Pre-warm media, trypsin–ethylenediaminetetraacetic acid (trypsin–EDTA), and phosphate buffered saline (PBS) without Ca^{2+} and Mg^{2+}.
3. Aspirate the media and discard.
4. Wash the cells with PBS.
5. Add trypsin–EDTA (approximately 1 ml per 25 cm^2).
6. Rock the flask to cover the monolayer.
7. Incubate the flask at 37 °C for 4 minutes.
8. Check under the microscope to make sure the cells are detaching from the surface and floating.
9. Gently tap on the sides of the flask to detach the remaining attached cells.
10. Add growth media (twice the volume as trypsin) to the flask and mix well.
11. Transfer the cell suspension to a centrifuge tube.
12. Centrifuge the tube at 200 × g for 5 minutes.
13. Take the supernatant out without disturbing the pellet.
14. Flick the tube to disturb the pellet.
15. Add 1 ml of fresh growth media and make an even suspension of cells by repeated pipetting.
16. Count the cells using the trypan blue protocol (below).
17. Add the required amount of growth media to the labelled flask.
18. Add the required amount of cell suspension to the flask and mix well.
19. Transfer the flask to the incubator.

Cryopreservation of cells is very important in cell culture. It enables a stock of cells to be stored for an extended period of time (e.g. many years). It reduces the risk of contamination, risk of phenotypic changes, cost and need to have the cell lines in culture all the time. It also enables experiments to be carried out using cells at a consistent passage number.

Cryopreservation of adherent cells

1. Prepare freezing media (10% dimethyl sulphoxide (DMSO) + 90% PBS).
2. Pre-warm freezing media, media, trpsin–EDTA, and PBS (without Ca^{2+} and Mg^{2+}).
3. Perform steps 3–14 of the sub-culture protocol.

4. Add 1 ml of freezing media and make an even suspension of cells by repeated pipetting.
5. Perform cell count and dilute the cell suspension in freezing media to achieve $2–4 \times 10^6$ cells per ml.
6. Add 1 ml of cell suspension into labelled ampoule (cryotube).
7. Place ampoule in a freezing flask (containing isopropanol) and leave it in a –80 °C freezer overnight (cells freeze at a rate of 1 °C/minute).
8. Transfer the ampoule to liquid nitrogen bank and record the position in the logbook.

Seeding cells on materials

For reproducible results it is important to determine the number of cells/cm^2 to be seeded on a material. Only by keeping the cell number constant from experiment to experiment can the biological material properties such as cell growth, attachment or the production of cellular factors be accurately assessed.

The trypan blue exclusion assay is used routinely to determine the seeding density. Trypan blue is a vital dye and does not interact with the cell unless the membrane is damaged. Cells that exclude the dye are viable. The proportion of viable and dead cells can then be determined with a haemocytometer.

Trypan blue staining of cells

1. Make a cell suspension according to the sub-culture protocol.
2. Prepare a haemocytometer and coverslip by spraying with 70% alcohol and drying with a clean tissue.
3. Fix a coverslip on the haemocytometer by moistening the coverslip with water or exhaled air, slide the coverslip over the chamber, and move back and forth exerting slight pressure until Newton's rings (rainbow-coloured rings) appear.
4. Place 0.2 ml of a suitable cell suspension (in complete medium) in a sealed container.
5. Add 0.2 ml of 0.4% trypan blue stain and mix thoroughly (dilution factor of 2).
6. Allow to stand for 2–3 minutes at 15 to 30 °C (room temperature). Prolonged exposure to trypan blue kills the cells.
7. With a pipette, fill both chambers of the haemocytometer. Do not over- or under-fill the chambers. Make sure that there are no air bubbles.
8. Under a microscope count the number of viable (unstained) and non-viable (stained blue) cells in eight–ten 4×4 squares or 0.1 cm^2 area (Fig. 20.2, CD Fig. 20.2). Count the cells in the squares and touching the left and top middle line.
9. Calculate the number of viable and non-viable cells/ml using the formulae given below.

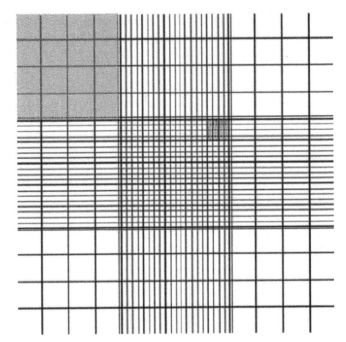

20.2 Diagrammatic illustration of a Neubauer haemocytometer chamber. The grey area highlights one 4×4 counting square (0.1 cm^2) (CD Fig. 20.20).

Number of viable cells/ml =	The average number of viable cells per 0.1 cm^2 area $\times 10^4$ (correction factor for volume of shaded area) $\times 2$ (dilution factor, 0.2 ml of cell suspension in 0.4 ml)

20.4 Basic techniques for assessment of cell viability

Assessing cell attachment and spreading on materials using scanning electron microscopy (SEM)

1. Grow the cells in flasks.
2. Follow the sub-culturing protocol to obtain the cells in suspension.
3. Count the cells using trypan blue protocol.
4. Decide on the density of cells needed to seed on the materials.
5. Seed the cells and allow them to grow for the desired period of time.
6. Make sure that you do not pipette or pour the solutions on top of the cells (there is a very high chance of washing the cells off).
7. Wash the cells with PBS twice.
8. Fix the cells with 2.5% glutaraldehyde in PBS for 40 minutes at 4 °C.
9. Rinse with PBS twice.

10. Dehydrate with graded series (increasing concentration) of ethanol.

(a) 25% – 5 minutes
(b) 50% – 5 minutes
(c) 70% – 5 minutes
(d) 90% – 5 minutes
(e) 100% – 5 minutes
(f) 100% – 5 minutes

11. Incubate in hexamethylsilasane (HMDS) for 5 minutes.
12. Incubate in fresh HMDS for 5 minutes.
13. Sputter coat samples with gold.
14. View under scanning electron microscope.

MTT [3-(4,5-dimethylthiazol-2-yl)-2,5-diphenyl tetrazolium bromide] assay

The MTT cell proliferation assay is a colorimetric assay system that measures the reduction of a tetrazolium component (MTT) into an insoluble formazan product by the mitochondria of viable cells. After incubation of the cells with the MTT reagent, a detergent solution is added to lyse the cells and solubilise the colour crystals. The samples are then read using an ELISA plate reader. The amount of colour produced is directly proportional to the number of viable mitochondria and therefore related to cell viability. The MTT system is a quantitative, more sensitive test than the trypan blue exclusion assay because there is a linear relationship between cell activity and absorbance; the growth or death rate of cells can be measured. The trypan blue test is qualitative and indicates only if a cell is alive.

The MTT assay protocol for adherent cells in 96 well plates is as follows.

1. Make a solution of 5 mg/ml MTT dissolved in PBS and filter sterilise.
2. 5 hours before the end of the incubation, add 20 µl of MTT solution from step one to each well containing cells.
3. Incubate the plate in a CO_2 incubator at 37 °C for 5 hours.
4. Remove media with needle and syringe.
5. Add 200 µl of DMSO to each well and pipette up and down to dissolve crystals.
6. Place plate into the 37 °C incubator for 5 minutes.
7. Transfer to plate reader and measure absorbance at 550 nm.

If larger plates are used, then adjust volumes accordingly.

20.5   Summary

Many of the advancements in understanding cell biology have arisen from cell isolation and *in vitro* studies. However, caution must be extended when comparing *in vitro* models to the multi-factorial and multi-cellular *in vivo* environment. Cells invariably behave differently *in vivo* due to the presence of other cell types, numerous cell signalling factors, the extracellular matrix and physiological differences in terms of mechanical stress, blood flow and 3-D growth. Cell type also has to be carefully selected for *in vitro* experiments to relate to the *in vivo* environment. For example, due to the ease of isolation, the vast majority of cell culture studies on endothelial cells have been performed on endothelial cells isolated from macro-vessels (large vessels), in human foreskin. However, it has now been shown that endothelial cell phenotypic expression and behaviour alters both between the type of vessel (blood, vein or lymphatic vessel), size of vessel (capillary, arteriole or artery) and also the location of the vessels (skin, brain, etc). Whether these differences in endothelial cell behaviour are induced by factors produced in the local environment and/ or due to different endothelial cell lineages is uncertain but clearly the selection of endothelial cell type has important consequences for the success of tissue-engineered tissues and biomaterial research.

20.6   Reading list

Davies J.M., *Basic Cell Culture*, Oxford University Press, 2002.

Freshney R., *Culture of Animal Cells: A Manual of Basic Technique*, New York, Alan R. Liss Inc., 1987.

Organisation for Economic Cooperation and Development, *Principles of Good Laboratory Practice*, Paris, OECD Publications, 1998.

Immunochemical techniques in tissue engineering and biomaterial science

GAVIN JELL and
JAMUNA SELVAKUMARAM
Imperial College London, UK

21.1 Introduction

Materials that are implanted into the body invariably cause a local (and possibly systemic) biological response. Traditionally, biomaterial implants were designed to limit the biological (and immune) response to the material, to create 'biological inertness'. More recently, research has been focussed on modifying the biological response by altering the material properties and thereby influencing the physiological environment, e.g. biodegradable or bioactive materials. In tissue engineering (TE) applications, understanding and optimising the cellular response to scaffolds (both prior and post-implantation) are essential for success (Chapters 18 and 19).

The biological response to biomaterials or tissue scaffolds is determined by a number of factors including the surface topography, the chemical composition, rate of resorption, type of dissolution product and mechanical properties. Materials are therefore developed with specific biological properties depending on the application. Some biomaterials are designed to promote cell-specific adhesion, whilst other biomaterials require anti-adhesive properties. For example, osseointegration (bone formation) surrounding orthopaedic implants is actively encouraged to prevent implant loosening (Chapter 13), i.e. with surface topography modifications or pro-adhesion/bioactive coatings, whilst other biomaterials such as vascular stents are developed to prevent protein and cell adhesion in order to avoid thrombus formation (blood clots) (Chapter 17). Considerable research is therefore concerned with monitoring cellular interactions with biomaterials and tissue scaffolds both *in vivo* and *in vitro*. Immunochemical techniques are the principal methods used to determine cell behaviour in response to biomaterials.

Immunochemical techniques enable the detection of proteins/molecules that possess a known physiological role in cells. For example, immunological techniques can detect proteins involved in cell attachment, cell activation, cell death and inflammation. Inflammation is a localised biological response elicited by injury or destruction of tissues, which serves to destroy, dilute or wall off (sequester) both the injurious agent and the injured tissue (Chapter 7).

Understanding these cellular processes is vital for biomaterial and tissue engineering research and design. Immunochemical techniques are extremely specific and sensitive, capable of detecting small differences between molecules in the presence of closely related contaminants, e.g. differentiating between macromolecules containing the L and D forms of a single amino acid in serum. Molecular biology techniques such as polymerase chain reaction (PCR), *in situ* hybridisation or Northern blotting are also commonly used to determine cellular behaviour to biomaterials by determining gene expression with mRNA detection.

One advantage of immunochemical techniques is that they can detect the final product of protein/molecule production, i.e. post-translation, rather than pre-translation mRNA detected by molecular biology techniques. Furthermore, immunochemical techniques also allow for the specific *in situ* location of intracellular or extracellular molecules.

21.2   Basic immunological principles

Whilst a basic understanding of immunology is desirable, it is not essential for the practical use of immunochemical techniques. The fundamental principle of immunochemical techniques is that a specific antibody (Fig. 21.1, CD Fig. 21.1) will bind to a specific antigen. An antigen is a foreign substance/molecule that elicits an immune response. An epitope is the antibody-binding site on the surface of an antigen; each antigen may contain one or more epitopes that can be recognised by a specific antibody.

Antibodies constitute a class of proteins called immunoglobulins (Igs), which are produced by B-cells (a particular type of white blood cell or lymphocyte). Immunoglobulins are Y-shaped molecules consisting of four polypeptide chains, each consisting of two identical light chains and heavy chains, linked together by disulphide bridges. Each chain has a constant region that is the same for all immunoglubulins of a certain class and a variable region that binds to specific antigenic epitopes. There are five types of constant regions that form the five classes of mammalian immunoglobulins, namely IgM, IgA, IgD, IgE and IgG. The antigen specific antibody is called the primary antibody. *In vivo* antibodies are important in the specific memory humoral immune response to infection.

Knowledge of the antibody preparation and target epitopes is important in interpreting immunochemical technique results. Polyclonal antisera (serum containing antibodies) are generated by the injection of a known antigen into an animal, which generates species-specific antibodies that are subsequently isolated. Polyclonal antisera include several different antibodies to the target protein and are therefore likely to target multiple isoforms of the target

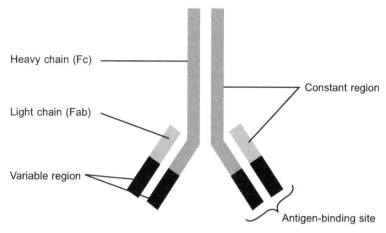

Heavy chain (Fc)

Light chain (Fab)

Variable region

Constant region

Antigen-binding site

21.1 Schematic representation of an antibody (CD Fig. 21.1).

antigen thus increasing sensitivity. However, the greater number of different antibodies to the target protein increases the likelihood of cross-reactivity with similar epitopes in other proteins and therefore increases the possibility of false-positive staining. Monoclonal antibodies are derived from the descendants of a single B-cell, thus producing identical antibody molecules. Monoclonal antibodies only identify one isoform of the target protein, therefore increasing the specificity but reducing sensitivity compared with polyclonal antibodies. A compromise therefore exists between sensitivity and specificity.

Most immunochemical techniques follow either a direct or indirect immunochemical procedure. In direct immunocytochemistry the primary antibody is labelled (normally with a fluorescent agent or a biotin molecule). In indirect immunochemistry a labelling agent binds to the unlabelled primary antibody. The labelling agent is commonly a species-specific labelled secondary antibody, e.g. fluorescent or biotin labelled mouse anti-human antibody (Fig. 21.2, CD Fig. 21.7).

Whilst indirect labelling acquires an additional step and therefore takes longer, there are several advantages over direct staining. Firstly, the primary antibody is normally in short supply and indirect immunochemical detection avoids the loss of primary antibody during labelling. Secondly, indirect labelling increases the sensitivity as several labelling agents can bind to the primary antibody and finally there are relatively few labelled primary antibodies commercially available.

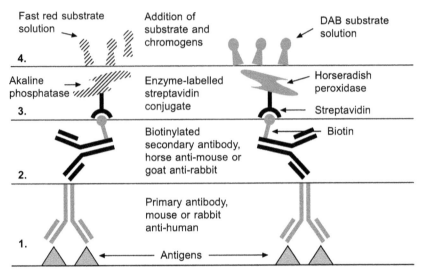

21.2 Indirect biotin–streptavidin immunocytochemistry. Primary antibody incubation (1) is followed by incubation with biotinylated secondary antibody raised against the primary antibody (2). The streptavidin–enzyme conjugate (streptavidin–HRP or streptavidin–AP) reacts with the biotinylated secondary antibody (3). This is followed by the addition of substrates and chromogens to form insoluble coloured products (4) (CD Fig. 21.7).

21.3   Common immunochemical techniques used in biomaterials

Immunochemical techniques are used to detect the expression of a particular antigen expressed either on the cell membrane, inside the cell, released by cell or in the extracellular matrix. Advantages and disadvantages exist for each immunochemical technique but the choice of technique depends largely upon the type of antigen expressed, the type of material used and the importance of qualitative versus quantitative data (Table 21.1).

For example, if the researcher is interested in the amount of macrophage colony stimulating factor (M-CSF, a factor important in osteoclast recruitment) released by bone cells grown on different materials, an enzyme-linked immunosorbant assay (ELISA) may offer the best immunochemical choice, as this allows quantifiable data on the amount of protein present in harvested cell culture media. The researcher must also consider that most biomaterials and tissue scaffolds are light impenetrable and therefore, to monitor cellular attachment/behaviour *in situ*, immunofluorescence or immunoelectron microscopy must be employed (Table 21.1). Alternatively, cells must be detached prior to analysis with Western blotting or FACS, which may affect cellular and extracellular protein expression or immunogenicity.

Table 21.1 The advantages and disadvantages of commonly used immunocytochemistry techniques in biomaterial and tissue engineering research

Immunocytochemical technique	Advantages	Disadvantages	Applications
Immunohisto-chemistry and immunocyto-chemistry	Specific localisation of antigens *in situ*. Staining lasts for years. Relatively inexpensive – no specialised equipment required	Not suitable for light impenetrable materials. Qualitative, and subjective without computer analysis	Commonly used for the examination of pathological samples, e.g. disease diagnosis in biopsies
Immunofluorescence	Has same advantages as immunohisto-chemistry but also enables the visualisation of cells grown on light impenetrable materials *in situ*. Can co-localise several antigens on the same cell	Fluorescence fades over time. Requires a relatively expensive fluorescent microscope. Some biomaterials auto-fluoresce	Widely used in cellular research both *in vivo* and *in vitro*
Western blotting	Sensitive, specific and semi-quantifiable depending on band intensity	Cells lysed prior to examination and therefore unable to localise cell specific antigen expression	Commonly used in cell culture research and tissue sections
ELISA	Quantifiable detection of antibody or antigen. Relatively quick and enables the detection of soluble factors released by cells. Very sensitive. Relatively cheap equipment	Often limited to commercially available ELISA kits. Does not allow *in situ* localisation of antigens	Frequently used for the detection of mediators released by cells in cell culture media and for diagnostic detection of antibodies to disease in human serum
Fluorescence activated cell sorter (FACs)	Allows the quantification of the proportion of cells expressing a certain antigen/antigens	Cells have to be detached from material prior to analysis. Requires expensive equipment	Commonly used in cell culture research and tissue sections
Immunoelectron microscopy	High magnification allowing sub-cellular location of antigens. Can be performed on light impenetrable materials	Expensive and time consuming, requires electron microscopy	Allows the intra-cellular location of antigens

Immunohistochemistry, immunocytochemistry and immunofluorescence

Immunohistochemistry, immunocytochemistry and immunofluorescence techniques (collectively termed immunostaining) detect and locate antigens expressed by cells either in tissue sections (immunohistochemistry) or in cell culture studies *in vitro* (immunocytochemistry). Immunofluorescence is performed both on tissue sections and *in vitro* cell cultures. To date, the most widely used immunostaining technique is indirect biotin–streptavidin single staining. This method is based on the extremely high affinity of streptavidin for biotin. The basic sequence of reagent application consists of the primary antibody, a biotinylated secondary antibody (of a different species to primary) and an enzyme-labelled streptavidin (normally conjugated alkaline phosphatase (AP) streptavidin or conjugated horseradish peroxidase (HRP) streptavidin), followed by addition of a substrate solution (Fig. 21.2, CD Fig. 21.7). Enzyme–substrate reactions convert colourless chromogens into coloured end products. This method is often termed the labelled streptavidin–biotin (LSAB) and has a greater sensitivity than other similar immunohistochemistry methods.

Immunostaining of cells *in vitro* follows the same method as immunohistochemistry apart from the fixing method, antibody incubation time and the greater care taken during washing to prevent cell detachment. There are several kits that amplify the signal response if there are few antigens or weak staining, e.g. the Envision© system (DAKO, Ely, UK). These systems often use an enzyme (HRP or AP)-labelled inert polymer 'spine' molecule that has secondary antibodies attached. The polymer contains an average of ten molecules of secondary antibody and 70 molecules of HRP enzyme. The increased numbers of enzyme amplify the intensity of staining and reduce incubation time.

Immunofluorescence has a key advantage in biomaterial research in that the investigator can visualise cells on light impenetrable materials (Table 21.1). Instead of a biotin-labelled secondary antibody employed in the LSAB method, immunofluorescence uses fluorescent-labelled primary or secondary antibodies. Immunofluorescence also enables the co-localisation of several antigens expressed on the same tissue or cells with different coloured fluorochromes (Fig. 21.3, CD Figs. 21.3, 21.4). Different primary antibodies raised in different animals (i.e. mouse- and rabbit-anti-human) can be mixed and incubated together followed by the addition of species-specific fluorescent-labelled secondary antibody. Prior to examination in a fluorescent or confocal fluorescent microscope, cells/tissues are commonly stained by a nuclear counter stain, e.g. DAPI. Confocal immunofluorescent microscopy is a relatively new technique, where a fine laser beam of light is used, and has several advantages over conventional fluorescence microscopy including clearer images and a controllable depth of field up to approximately 100 μm.

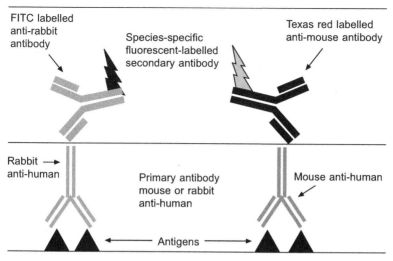

Rabbit → anti-human

Mouse anti-human

Primary antibody mouse or rabbit anti-human

Antigens

FITC labelled anti-rabbit antibody

Species-specific fluorescent-labelled secondary antibody

Texas red labelled anti-mouse antibody

21.3 Schematic of double indirect immunofluorescence. Primary anti-human antibodies for two different antigens, each from a different species (i.e. mouse- and rabbit-anti-human antibodies) were incubated together. Following washing, species-specific fluorescent-labelled secondary antibodies, differing in their fluorescent conjugates (i.e. horse anti-rabbit FITC and horse anti-mouse Texas red labelled antibody) were mixed and incubated together.

Results of immunostaining can be interpreted as the proportion of positively stained cells, the distribution of the staining and the intensity of staining. The number of positive cells can be counted in randomly selected microscope fields. However, the percentage of positive cells in a specified area does not allow for interpretation of staining intensity. An increased number of epitopes expressed by a cell would bind more antibodies and hence increase the staining intensity. Therefore, it is sometimes possible to interpret semi-quantitatively both the proportion of positive cells and intensity of staining. The percentage grade of positive cells can be multiplied by the intensity grade to give a staining index. To avoid bias and subjective interpretation of immunohistochemistry results investigators should evaluate samples blindly and independently. Computer-aided image analysis also reduces subjectiveness and allows for quantification of immunostaining.

Western blotting

In the Western blotting technique, proteins in homogenised cells or tissues are separated electrophoretically on sodium dodecylsulphate–polyacrylamide gel electrophoresis (SDS-PAGE) gels according to their molecular weight (MW); proteins are then transferred to nitrocellulose or PrDF membrane where antigens are immunochemically detected by specific antibodies as

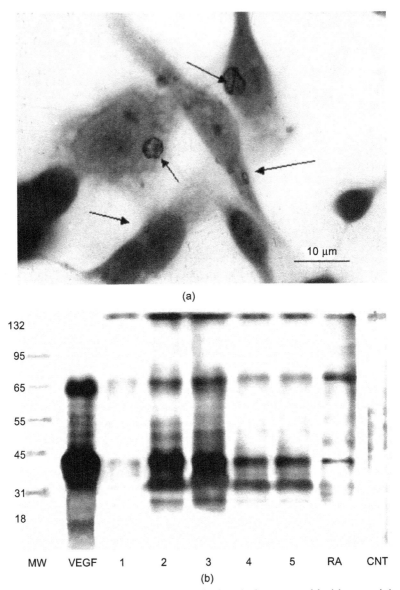

21.4 Examples of immunochemical techniques used in biomaterial research. (a) Vascular endothelial growth factor (VEGF) expression by indirect biotin–streptavidin single staining on endothelial cells exposed to hydroxylapatite particles *in vitro.* Note attached particles (arrows). VEGF (molecule weight 45 kD) expression is also shown in the digested tissue surrounding failed orthopaedic implants in a Western blot. (b) The first lane contains labelled proteins of a known molecular weight (MW), in the second lane recombinant VEGF protein is used as a positive control, the remainder of the lanes are the proteins digested from the experimental samples apart from the negative control where the primary antibody (anti-human VEGF) was omitted (CNT).

described for immunocytochemistry (Fig. 21.4). Cells grown on biomaterials or tissue constructs *in vitro* have to be removed prior to homogenisation and protein separation. Cell-specific antigen expression in the local *in situ* environment cannot therefore be observed; for example, osteoblasts may have altered phenotypic expression on different surfaces within a tissue scaffold. However, the homogenisation process used in Western blotting may reveal antigens that are concealed in immunocytochemistry due to fixing methods.

To ensure the reproducibility of results and allow the comparison of band intensities from different cell populations or tissues, the protein concentration of the homogenised cells/tissue has to be calculated prior to electrophoresis. Protein concentrations are typically determined using an assay based on the reaction of protein with an alkaline copper tartrate solution and Folin reagent. To estimate protein size in Western blots, a MW marker is used that contains distinct biotinylated protein bands within a molecular weight range of the antigen of interest.

Enzyme-linked immunosorbent assay (ELISA)

ELISA allows for precise quantification of the amount of antigen and antibody present in a sample, which is particularly useful when comparing cellular responses to different biomaterials or tissue scaffolds. For example, the inflammatory response to different orthopaedic biomaterial wear particles could be measured by the relative amount of pro-inflammatory or anti-inflammatory cytokines released by cells *in vitro* or in the serum levels of animal models.

There are three main types of ELISA: an indirect ELISA where an antigen-coated plate measures the amount of antibody, a sandwich ELISA where an antibody-coated plate measures the amount of antigen, and a competitive ELISA, where the concentration of antigen is inversely proportional to the colour produced. Pre-prepared standards (often recombinant proteins) and samples are added to the antigen or antibody-coated well plate, where any specific antibody–antigen binding will be immobilised at the plate. After washing, an enzyme-linked antibody is added to the wells, followed by the addition of the substrate solution. The colour develops proportionally to the amount of antigen/antibody present. The colour intensity is measured with a plate reader. Quantifiable results are calculated from the standard curve created from serial dilutions of the standard/recombinant protein.

Fluorescence activated cell sorter (FACS)

Flow cytometry or fluorescence activated cell sorting (FACS) allows for the quantification of the percentage of cells that are positive for particular antigens in a cell population. Following a similar method to direct or indirect immunofluorescence, antigens are labelled with fluorochromes. However, unlike immunofluorescence, single cells flow past an excitation source, positive cells absorb the light and re-emit fluorescence, which is measured by a detector. The proportion of positive cells in the population is then calculated. For flow cytometry, cells that are need to be in suspension and therefore tissue sections have to be degraded and adherent cells that are attached to biomaterials must be detached. Localisation of antigens both within a cell and cellular expression in their local environment (*in situ*) is therefore not possible. Flow cytometry also detects non-reflective light and can therefore measure the number of cells with internalised or adhered particles.

Immunoelectron microscopy

This technique allows the investigator to visualise cells grown on light impenetrable surfaces at high resolution The antibody/antigen complexes can be localised to a particular sub-cellular organelle by using antigen/antibody gold labelling. Following primary antibody incubation, cells or tissue are then incubated with protein-coated gold particles (size range is 5 to 20 nm). The gold particles bind to the Fc portion of the antibody and are detected by electron microscopy. Scanning electron microscopy also allows imaging of cell morphology and attachment. The number and length of filopodia (cell protrusions) can be used as a measure of cell attachment.

Controls

Controls are necessary to validate the immunochemical results. Negative controls check for non-specific binding and include the omission of primary antibodies, or substitution of primary antibodies with non-immune immunoglobulin from the same species. Positive control of antibodies to a known antigen present in cells/tissue confirm a correctly working staining procedure.

Non-invasive techniques

Whilst immunochemical techniques are extremely specific and sensitive, they are invasive and cannot be performed on live cells *in situ*. Novel non-invasive methods to assess live cell behaviour *in situ* on light impenetrable surfaces are required. This is of particular importance in tissue scaffolds and cell-seeded implants, i.e. vascular grafts, where cell behaviour can be monitored prior to implantation. Raman spectroscopy may offer a novel solution whereby a biochemical spectral fingerprint of individual cells is obtained and is currently under investigation.

Table 21.2 Examples of immunochemical markers of cell behaviour used in biomaterial research

Biological response	Biomaterial research applications	Specific immunochemical markers
Inflammation	Inflammatory response to orthopaedic wear debris is believed to be the primary reason for orthopaedic implant failure	Pro-inflammatory cytokines (IL-1, TNF-α). Anti-inflammatory cytokines (IL-10)
Cell attachment	Frequently measured to determine substrate biocompatibility *in vitro* for a variety of biomaterials and tissue scaffolds. Materials may be seeded with proteins to either aid or inhibit cellular attachment.	Integrins, e.g. $\alpha v \beta 3$, vitronectin, focal adhesion kinase (FAK)
Cell death (apoptosis or necrosis)	Toxic leaching or other material characteristics may cause cell death. Apoptosis (programmed cell death) causes a significantly different biological response to necrosis	p53, PARP, BAX, caspases
Cell stress	To assess the health of cells exposed to biomaterials, cell stress can be analysed	Heat shock proteins (HSPs), are expressed by cells undergoing a variety of environmental stresses
Cell activation	Biomaterials may affect the cell cycle, either stimulating or inhibiting cell division. Cell cycle stage influences cell phenotypic responses	Ki-67 is expressed by cells that are in active stages of their cell cycle (i.e. not in G0)
Bone formation	Considerable research is focused on generating new bone formation, different materials affect bone nodule formation *in vitro* and *in vivo*	Alkaline phosphatase, osteocalcin, bone morphogenic protein–1 (BMP-1), collagen
Angiogenesis (blood vessel formation)	Angiogenesis is vital for both bone formation and inflammation and therefore has numerous applications in biomaterials and tissue engineering	Pro-angiogeneic vascular endothelial growth factor (VEGF). Angiogeneic inhibitors endothelin, angiostatin

21.4   Immunochemical applications in biomaterial science and tissue engineering research

Immunochemical techniques are used to determine cellular behaviour in numerous biomaterial and tissue engineering applications (Table 21.2). For example, cell adhesion to biomaterials is most commonly used as a measure of substrate biocompatibility *in vitro*. The process of cellular attachment to substrate initially involves serum proteins such as fibronectin, vitronectin, or fibrinogen (present in culture media), which attach to the material surface. Cells then bind to these attached proteins using cell–matrix adhesion proteins.

The properties of a material may influence cells directly, or indirectly depending upon the nature and orientation of adsorbed proteins prior to cell attachment. Important factors that determine protein and cellular adhesion to materials include chemical composition and surface topography. For example, the biological performance of implantable titanium devices in medicine and dentistry depends critically on their micro- (structures larger than 1 μm) and nano- (structures smaller than 1 μm) topography. Osteoblast cells also react differently to different diameter pore sizes in tissue scaffolds. Surface physicochemical properties also play an important role in cell and protein interaction with materials. These include surface energy, hydrophilic or hydrophobic nature, surface charges and reactive surface groups.

The interaction of cells with a surface can be greatly influenced by the matrix proteins because of the presence of glycoproteins on cell membranes called integrins. Integrins are extracellular matrix receptors that have selective affinity for certain matrix proteins to which they bind with relatively low affinity. The low affinity calcium- and magnesium-dependent binding provides a Velcro® type effect that allows the cells to explore their environment. The binding involves an extracellular binding site and an intracellular site, which in many cases is linked to structural proteins, e.g. actin filaments, and intermediate filaments. Integrins have alpha and beta subunits; depending on the subunits, they bind to specific matrix proteins. For example, cells containing integrins with alpha-6 and beta-4 subunits bind to matrix proteins in the basal lamina, e.g. laminins, and intermediate filaments, e.g. keratins, in the cytoplasm. On the other hand, cells containing integrins with alpha-5 and beta-1 bind to fibronectin in the extracellular matrix.

Whilst this chapter has concentrated on *in vitro* biomaterial research, immunochemical techniques are also frequently used to understand complex biomaterial–cellular interactions *in vivo*. For example, orthopaedic bone–implant-interface tissue is analysed frequently by immunochemical techniques to understand the complex multifactorial and multicellular process involved in implant osseointegration and rejection that are impossible to model accurately *in vitro*.

21.5   Summary

Immunochemical techniques are a powerful, specific and sensitive tool for understanding cellular response to materials. However, care must be taken in immunochemical experiment design and interpretation to ensure truly representative and subjective results. This chapter has outlined the basic immunochemical techniques and described some immunochemically detected markers of cellular behaviour important in tissue engineering and biomaterial research.

21.6   Key definitions

Antibody: an antigen-binding immunoglobulin produced by b lymphocytes.

Antigen: a foreign substance/molecule that elicits an immune response.

Apoptosis: programmed cell death.

Confocal microscopy: a system of (usually) epifluorescence light microscopy in which a fine laser beam of light is scanned over the object through the objective lens.

Epitope: an antibody binding site on an antigen, also called an antigenic determinant.

Inflammation: a localised biological response elicited by injury or destruction of tissues, which serves to destroy, dilute or wall off (sequester) both the injurious agent and the injured tissue.

In situ: in the natural environment.

In vitro: describes biological phenomena that are made to occur outside the living body.

In vivo: describes biological phenomena that occur or are observed within the bodies of living organisms.

21.7   Reading list

Alberts B., *Molecular Biology of the Cell*, 4th edition, New York/London, Garland Science, 2002.
Benjamin E., *Immunology: A Short Course and Student Guide*, Chichester, John Wiley and Sons, 2001.
Clancy J., *Basic Concepts in Immunology; A Student's Survival Guide*, New York, McGraw-Hill, 1998.
Hay F. and Westwood O., *Practical Immunology*, Oxford, Blackwell Science, 2002.
Kuby J., *Immunology*, 3rd edition, New York, W.H. Freeman, 2003.

Lanza R., Langer R. and Chick W.L., *Principles of Tissue Engineering*, Philadelphia, Academic Press, 2000.

Pound J., *Immunochemical Protocols*, New Jersey, Hamana Press, 1998.

Schantz M., *From Genes to Genomes: Concepts and Applications of DNA Technology*, Chichester, John Wiley & Sons, 2002.

William P., *Fundamental Immunology*, 4th edition, Philadelphia, Lippincott-Raven, 1999.

22

Clinical applications of tissue engineering

LARRY L. HENCH and
JULIAN R. JONES
Imperial College London, UK

22.1   Introduction

The concepts involved in growing tissues and organs outside the body followed by implantation (tissue engineering) or using cells to stimulate repair *in vivo* (regenerative medicine) have been described in Chapters 18–21 along with the materials and methodology required. This new field offers the potential for solving the problem of shortage of organ donors needed to maintain quality of life for an ageing worldwide population. The objective of this chapter is to summarise the current status of clinical use of tissue-engineered constructs. Most tissue-engineered applications are still experimental with limited human trials. The CD lecture for Chapter 22 provides details and an update.

Several technological barriers must be overcome in order for clinical use of tissue-engineered constructs to become routine.

Cell source

Chapter 18 describes the options available for cells to use in tissue-engineered constructs, including stem cells. At present, the only reliable cell source is autologous cells from the patient. This source has serious limitations of numbers of cells available and ability of the cells to maintain a phenotype capable of generating extracellular proteins. Cloned, immortal cell lines are capable of proliferating but usually lack the differentiation needed for stable tissue repair. See Chapter 13 by D.K. Buttery and J.M. Polak, in *Future Strategies for Tissue and Organ Replacement* in the Reading list for details.

Stable 3-D constructs

All tissues and organs have a complex interdependence of cell types with an interconnected 3-D architecture. Most tissue-engineered constructs involve only one, or at most two, cell types grown primarily in a 2-D configuration. This compromise in structure limits the clinical viability of the constructs.

Vascularisation

All tissues and organs have an interpenetrating network of blood vessels connected to the circulatory system to provide nutrition and eliminate waste products. Tissue-engineered constructs at present lack this vital network when they are transplanted. The host tissues must quickly infiltrate the tissue-engineered graft with a blood supply or the cells will die. One of the major challenges of tissue engineering is to achieve angiogenesis rapidly after implantation.

Interfacial stability

The limitations of tissue-engineered constructs listed above often result in problems at the host tissue–graft interface. Shrinkage, infiltration by new tissue or breakdown leads to less than desirable clinical outcomes.

Sterilisation

Maintaining sterility of a tissue-engineered construct that contains living cells is a serious challenge in manufacturing, handling, storage and transport. Most methods used for sterilisation of non-living implants and devices, such as γ-irradiation, autoclaving, ethylene oxide and UV light will kill cells. Sterility must be achieved during processing and maintained until implantation is complete.

Cost

All of the above factors add to the manufacturing costs and presently limit many tissue engineering applications to exploratory patients.

Survivability

Long-term survivability of tissue-engineered constructs is uncertain. Consequently, in many cases use is restricted to applications where no other procedure is available, as required by ethical considerations (Chapter 25). These 'worse case' scenarios make it difficult to assess viability and success of the new procedures.

Regulatory considerations

Tissue-engineered products are subject to the same regulatory procedures as non-living biomaterials and devices (Chapter 24). At present, only a few products have been produced to meet these regulatory requirements. Costs and risk/benefit factors are hard to predict because of the uncertainty of regulatory approval.

22.2   Skin

Many patients need a supply of replacement skin to treat severe burns, loss of skin in accidents, chronic ulcerations, especially in the legs and feet, and for repair of sites where tumours have been removed. The clinical 'gold standard' is use of the patient's own skin in the form of a split-thickness autograft. The advantage is the absence of immunorejection but the disadvantages are limited availability of normal skin, morbidity of the site of the autograft and final cosmetic outcome. Tissue engineering of skin substitutes offer a much needed solution to these problems. The objective is to produce a 'ready-to-use' skin substitute that performs as well as the patient's own skin. Developments for more than 20 years show promise, but as Otto has reviewed in Chapter 4 of *Future Strategies of Tissue Engineering and Organ Replacement* (see Reading list), *'It remains to be seen which commercially available skin substrates will be successful over the long term, particularly as far as the patient is concerned.'*

Integra® is a dermal substitute material that has been used to treat skin wounds for more than a decade. It consists of a bovine collagen-derived membrane with cross-linked chondroitin-6-sulphate bonds covered with a removeable layer of polysiloxane. The dermal substitute is placed on the debrided site of the wound and, after the early stages of wound healing and vascularisation, a surface epidermal layer is applied, either as a split thickness autograft or as cultured keratinocyte sheets. Results are generally favourable for burns patients with an otherwise poor prognosis.

Other clinically important tissue-engineered products used as skin substitutes include Alloderm®, Apligraf®, Dermograft®, Transcyte® and Laserskin®. See Chapter 4 by Otto, in *Future Strategies for Tissue Engineering and Organ Replacement*, in the Reading list for details of these skin substitutes, their relative success clinically and an extensive reference list. The review by I. V. Yannas (Chapter 134 in the *Biomedical Engineering Handbook*) also provides an excellent overview of the use of artificial skin and dermal substitutes, including clinical results.

22.3   Cartilage

Articulating cartilage does not have a blood supply so has little capacity for self-repair (Chapter 8). Tissue-engineered repair of defective cartilage caused by arthritis or trauma could potentially benefit an estimated 1 million patients per year. Animal studies for 30 years have shown that transplantation of isolated chondrocytes can enhance articular repair, as reviewed by Freed and

Vunjak-Novakovic (Chapter 120 in the *Biomedical Engineering Handbook*) in the Reading list.

Three tissue-engineered options are available clinically after harvesting chondrocytes from a patient: (1) expand the cells to a population of millions per cm^2 *in vitro* followed by injection into the site of damage, (2) grow an expanding cell population on a porous resorbable fibrous scaffold, such as an 85:15 combination of PLA/PGA, followed by implantation into the defect site, or (3) harvest chrondrocytes as small plugs and transplant them into the defect creating a 'mosaicplasty'.

Long-term survivability is a concern for all three approaches due to difficulty of the chrondrocytes in maintaining a differentiated phenotype in culture. The complex 3-D architecture of articulating cartilage described in Chapter 8, and the extracellular matrix required to provide proper load transfer to the cells, is seldom achieved. Interfacial stability between the host cartilage site and the tissue-engineered construct is also a common problem. At present, relatively few patients have had a tissue-engineered cartilage repair.

22.4   Tendons, ligaments and bone

Chapters on bone and cartilage reconstruction by the pioneering tissue-engineering team of C. A. Vacanti and J. P. Vacanti, and tendons and ligaments by Goulet *et al.* in the Lanza, Langer and Chick book '*Principles of Tissue Engineering*' describe the progress being made in the use of tissue-engineered approaches to replace these vital components of the musculoskeletal system. In general, there is still much research required to achieve the ultrastructural characteristics of the cell–extracellular matrix of the natural tissues. The poor mechanical properties of the tissue-engineered constructs are a consequence, in part, of the difficulty of applying proper mechanical–chemical stimuli to the cells as they proliferate *in vitro* in their tissue-engineered scaffolds. Current research into the use of bioreactors that can apply such stimuli in a controlled manner may solve this problem. In the meantime there are few clinical applications that have had long-term success.

22.5   Pancreas (islets of Langerhans)

More than 100 million people worldwide suffer, in varying degrees, from diabetes mellitus. Daily insulin injections are commonly used to treat this condition but there is growing morbidity of the complications associated with the disease, including: large-blood-vessel pathology and neuropathies of the autonomic, peripheral and central nervous system. Blindness, amputation and end-stage renal failure are all too often long-term consequences of diabetes.

Thus, there is impetus for tissue engineering to provide a solution to the production of insulin *in vivo* when the pancreas is unable to do so. Several approaches to producing an 'artificial pancreas' have been investigated for the last 30 years, as reviewed by Galletti *et al.* in Chapter 130 of the *Biomedical Engineering Handbook*. Few are clinically important as yet. One of the most promising is to harvest and grow pancreatic islets of Langerhans cells *in vitro*, encapsulate or immobilise the cells with polymeric scaffolds and transplant the tissue-engineered constructs. The chapter by Lanza and Chick in *Principles of Tissue Engineering* describes three major types of encapsulation systems investigated. Animal studies show that these tissue-engineered systems can function for several months to more than a year. The data also show that a major hurdle is to maintain function of the islets over prolonged periods of time.

22.6   Liver

More than a million people worldwide require hospitalisation for liver diseases; more than 25 000 patients die annually in the USA from chronic liver disease. A limited donor supply makes liver transplantation a solution for only a few thousand patients. A tissue-engineered solution is a possibility because the liver has a large potential for self-repair. Thus, a short-term use of a 'liver support' device that provides time for regeneration of normal liver tissue could save thousands of lives annually and be highly cost effective.

Two approaches are being investigated: (1) transplantation of hepatocytes by direct injection and (2) *in vitro* growth of hepatocytes on resorbable polymer scaffolds, such as surface-modified PLA/PGA or alginate beads, to create a tissue-engineered construct that could either be transplanted or used for blood waste removal by shunting a portion of the blood through the device.

As with most tissue-engineered approaches to organ engineering, the fundamental problem is maintaining the complex, multiple biological functions of hepatocytes for sufficient time to allow the liver to repair. See Chapter 6 by Selden and Hodgson, in the book *Future Strategies for Tissue Engineering and Organ Replacement* and Chapter 117 by Kim and Vacanti in *The Biomedical Engineering Handbook* for details. Clinical use of a tissue-engineered liver is likely to be 5–10 years away. As Kim and Vacanti summarise, '*Currently, the amount of hepatocyte engraftment, proliferation, and the duration of hepatocyte survival remain undetermined.*'

22.7   Kidney

Replacement of the filtration function of the kidney with haemodialysis or chronic ambulatory peritoneal dialysis is described in Chapter 17. Although this is one of the most successful uses of an 'artificial organ', it is restricted to a filtration function. Dialysis does not replace the complex combination of regulatory, metabolic, endocrine or homeostatic functions of the kidney. Consequently, dialysis patients still have significant health complications. A tissue-engineered biohybrid kidney would help eliminate these complications. Chapter 121 by Humes in the *Handbook of Biomedical Engineering* summarises the approach needed to replace the critical elements of renal function, including excretory, regulatory (reabsorptive) and endocrinologic functions.

Two main units are needed to replace the renal excretory function: (1) the glomerulus, which provided filtration, and (2) the tubule, the readsorber. Humes' chapter describes progress in developing a bioartifical glomerulus, a bioartificial tubule and a bioartificial endocrine gland that acts as an erythropoietin generator. A functioning tubule will probably require the use of stem cells. The economic benefits of success are projected to be enormous, in excess of $10 billion per year. The barriers to achieving this success are the same as described above: the ability to achieve and maintain a 3-D architecture of highly differentiated cells for long periods of time.

22.8   The cardiovascular system

Chapter 9 describes the high mortality resulting from disease of the cardiovascular systems and Chapters 14 and 17 discuss the use of various devices to prolong life. For example, replacement of heart valves is routine, saving more than 100 000 lives annually, using either mechanical prostheses or tissue valves. Commercially available tissue valves include heterograft (porcine and bovine, cross-linked with gluteraldehyde) and autograft. The disadvantage of all tissue valves is limited durability (10–15 years), as reviewed by Love in Chapter 28 in *Principles of Tissue Engineering* and by Schoen *et al.* in *Cardiovascular Pathology* (see Reading list). Use of autologous tissue, such as the pericardium, for valve repair or replacement is a step towards a tissue engineering solution. The technical issues are summarised by Love and early clinical success is documented.

Growth of heart valves *in vitro* using stem cell technology is the goal of several laboratories. Success would be a major breakthrough in treating valve disease.

Vascular replacement or bypass is used to prolong the life of hundreds of thousands of patients annually. Only large- or medium-sized arteries can be used due to limitations of graft function for current implant materials, as summarised in Chapter 17 and discussed by Zarge *et al.* in their chapter on blood vessels in *Principles of Tissue Engineering*. Efforts to grow microvascular blood vessels *in vitro* show promise by using fibroblast growth factors (FGF) to induce capillary endothelial cells to invade 3-D collagen or fibrin matrices and form tubules that are similar structurally to blood vessels. Control of the micro-environment of cells to promote natural angiogenesis offers promise for producing tissue-engineered constructs suitable for vascular grafts in the near future. See the Nerem, Matsumura *et al.*, and Neumenschwander and Hoestrup articles in the Reading list for an update.

22.9   Nerves

The peripheral nervous system has the potential for self-repair following injury but clinical success is variable, as reviewed by Valentini in Chapter 3 in the *Biomedical Engineering Handbook* and by Valentini and Aebischer in *Principles of Tissue Engineering*. Fusing techniques, such as using fibrin glue or polyethylene glycol (PEG) have been tried but the most promising approach is to use nerve guidance channels. Surface modification of various polymeric systems enhances the migration and importantly connects the nerves growing along the channels. Use of third-generation biomaterials that provide controlled release of neurotrophic factors that are specific to nerve outgrowth, neuronal survival, and nerve sprouting appears to be the key to achieving a tissue-engineered solution (see the Hench and Polak article in the Reading list).

22.10   Summary

Tissue engineering of organs shows promise at a level of basic research. Many limitations associated with achieving and maintaining large 3-D constructs of cells that are fully differentiated restrict clinical use. Repair of skin and cartilage with tissue-engineered approaches is clinically available, although problems of long-term stability still exist. Limited clinical trials are just beginning in other applications.

22.11   Reading list

Borozino J.D. (ed), *The Biomedical Engineering Handbook* (Chapter 3, 117, 120, 121, 130), Boca Raton, Florida, CRC Press, 1995.

Hench L.L. and Polak J.M., Third generation biomedical materials, *Science*, vol. 295 (5557), p. 1014, 2002.

Lanza R., Langer R. and Chick W.L. (eds), *Principles of Tissue Engineering*, London, Academic Press, 2000.

Matsumura G., Hibino N., Ikada Y., Kurosawa H. and Shinoka T., Successful application of tissue engineered vascular autografts: clinical experience. *Biomaterials*, vol. 24, p. 2303, 2003.

Nerem R.M., *Critical Issues in Vascular Tissue Engineering, Int. Congress Series,* vol. 1262, p. 359, 204.

Neumenschwander S. and Hoestrup S.P., Heart valve tissue engineering, *Transplant Immunology*, vol. 12, p. 359, 2004.

Polak J.M., Hench L.L. and Kemp R. *Future Strategies for Tissue Engineering and Organ Replacement* (Chapters 4–8, 15), London, Imperial College Press, 2002.

Schoen F. *et al.*, in *Cardiovascular Pathology*, vol. 1, no. 1, p. 29–52, 1992.

Part V

Societal, regulatory and ethical issues

23

Regulatory classification of biomaterials and medical devices

DAVID C. GREENSPAN
US Biomaterials, USA

LARRY L. HENCH
Imperial College London, UK

23.1 Introduction

The classification of medical devices for regulatory purposes is related to the inherent risks of the device. Different regulatory control mechanisms are assigned to each class. In the USA there is a three-tiered system. Class 1 devices possess the lowest risk to the patient. Class 2 devices require more rigorous scrutiny and control in order to gaining regulatory clearance. Class 3 devices require the most stringent controls and manufacturing reporting requirements. There are two main paths to gaining regulatory status in the USA: clearance by demonstrating that the device to be marketed is substantially equivalent to a device marketed prior to enactment of the 1976 Medical Device Amendment (known as a 510(k) clearance), or a Pre-Market Approval which entails the submission and review of detailed data pertaining to the device to verify safety and efficacy (21CFR 822.16).

In Europe there is a four-tiered system based on the degree of risk associated with device usage, the amount of time the device is in contact with the human body, and the degree of invasiveness of the device (Medical Device Directive EU 93/42). Medical devices are part of a harmonised system of quality and safety legislation for all products sold in Europe. Products that conform to these regulations bear the 'CE Mark'. The European Commission issues directives to co-ordinate the system throughout the European Union. It is these directives related to medical devices that specify the classification and other essential requirements to be met by manufacturers and member states. These basically define the quality and safety requirements of the medical devices. The directives also specify the processes that must be followed by those involved in the manufacture, sale, use and regulation of medical devices.

This chapter is divided into two parts. The first part is a brief summary of the major concepts governing the regulatory situation for medical devices both in the USA and in the European Union (EU). The second part, presented in the CD lecture, demonstrates how the regulatory system performs through a hypothetical case of a company developing a new medical device.

Medical device

A medical device is an instrument, apparatus, implement, machine, contrivance, implant, *in vitro* reagent, or other similar or related article, including any component, or part that involves:

- diagnosis, cure, mitigation, treatment or prevention of a disease or condition;
- the structure or function of the body;
- non-achievement of intended use through chemical action;
- non-dependence upon being metabolized.

The above definition has been adopted by both the EU and the Food and Drug Administration (FDA) in the USA. A medical device, importantly, does not achieve the intended use through chemical action and is not dependent on being metabolised. These last two characteristics relate to drugs and pharmacological compounds and this is what differentiates drugs and pharmacologic agents from medical devices.

23.2   How are medical devices regulated?

There are a number of levels of regulation for various medical devices. In the USA, by an Act of the Congress, the FDA is the authority that regulates and controls the sale and distribution of medical devices. In the UK, it is the Medical Devices Agency. In Japan, it is the Ministry of Health and Welfare. Most countries in the world have some form of agency to control the sale and distribution of medical devices. In addition, the EU has created, through Treaties, a system for control and regulation of medical devices involving conformity marketing, the CE mark.

Medical devices are regulated by various authorities
- USA – Food and Drug Administration (FDA)
- UK – Medical Devices Agency
- Japan – Ministry of Health and Welfare
- European Union – CE Marking

23.3   Classification of medical devices

The definition of medical devices encompasses an enormous number of products that range from *in vitro* reagents for diagnostics to simple devices, such as temporary skin dressings, to very complex, life-sustaining devices such as pacemakers and heart valves. Therefore, there is a need to differentiate the risks associated with these devices and this has led to a classification system. The FDA classifies medical devices by the risk associated with using that device for the patient. Class 1 devices generally are low risk and expose the patient to little danger, such as *in vitro* reagents for testing for diseases or certain types of conditions, or non-invasive products like bandages. Class 2 devices pose a higher risk to the patient and may include non-invasive or invasive, short-term devices, like catheters that are only placed for a short term, or some of the electronic devices that are used to test or to analyse a patient's health. EKG (electro cardiogram) and radiography equipment are Class 2 devices.

FDA device classifications
- Class 1 – devices are generally low risk and non-invasive
- Class 2 – medium risk and include both non-invasive and invasive (short-term) devices
- Class 3 – generally high-risk, long-term implantable devices

In the EU, Class 2a devices are generally very similar to Class 2 devices as assigned by the FDA in the USA. However, the EU has created a Class 2b, which are implantable devices that include resorbable materials and other types of long-term implants. Class 3 devices are generally associated with a high risk to the patient should they fail and are generally long-term implantable devices that are either considered to be life-sustaining or, should they fail, would result in severe compromise to the patient's health and well-being. Class 3 devices include pacemakers, total hip and knee replacements, bone graft materials, cardiac assist devices and heart valves.

23.4 History of regulatory agencies

The USA FDA was created by an Act of the United States Congress in 1906 through the passage of the Food and Drug Act. The main purpose of this bill was to prohibit what was, at the time, widespread misbranding and adulteration of foods, drink and drugs. The bill predominantly focussed on adulteration of food and drink and, to a lesser degree, the misbranding of drugs. In the early days it was aimed at protecting the consumer in those areas. Over the years, the USA Congress has continually revised and refined the breadth and the scope of the enforcement and oversight of the FDA.

In 1938 the Congress passed the Food, Drug and Cosmetic Act, which set up and gave specific guidelines for what cosmetic and drug manufacturers were required to do to prevent the misbranding and adulteration of their devices. Prior to this, laws were rather vague and were generally ineffectual. The Act in 1938 established the role of the FDA in overseeing and allowing the sale and distribution of drugs and cosmetics and established their authority to prosecute companies acting outside of the law. In 1976 a critical piece of legislation was passed known as the Medical Device Amendments. This established, in the USA, clearer definitions as to what medical devices were, defined the classification process, established ways for the medical device manufacturers to get products into the market and created a review process for products that were already in the market in order to ensure that the manufacturers were making devices that were both safe and efficacious for use. It is this 1976 amendment that has resulted in much of the present regulation within the medical device industry. This was clearly a landmark bill passed by Congress. In 1990, the Congress went further in enacting the Safe Medical Devices Act. This came about after a review of the 14 years of experience since the passage of the 1976 Medical Devices Act.

The 1990 Act increased control of the FDA by allowing the agency to require companies to submit performance data so that the FDA could better assess performance of devices. It placed a higher burden on manufacturers to ensure that their medical devices were performing as intended in a safe and effective manner. It also required that medical device manufacturers be able to track their devices after they leave the factory. Manufacturers of Class 3 high-risk devices are now required to ensure a method of tracking a device from the start of manufacturing to implantation. In this way, if something goes wrong with a device after implantation it can be tracked to the patient and, if possible, retrieved and changes made. This revision in the medical device legislation also reclassified many of the devices that had been Class 3 devices to Class 2 devices.

In 1997, as a result of the EU issuing the Medical Device Directives, and with the recognition that the two largest markets for medical devices were operating under two different systems, Congress passed the 1997 FDA Modernisation Act. The main thrust of this Act is to move FDA rules and regulations towards a global harmonisation with other nations so that medical device regulations can be standardised throughout the world. It also modified guidelines to make the regulations in place by the FDA clearer for the manufacturers of medical devices. With this Act, the manufacturers know more exactly what is required in order to have a device cleared for sale by the FDA. It also clarified one of the major pathways for clearance of medical devices, i.e. the 510(k) pre-market notification route. The '510(k)' designation refers to a specific paragraph in the original medical device legislation of 1976. This section is the most widely used method of obtaining FDA clearance to sell a medical device in the USA.

With the creation of the EU in the 1980s, there was a move to allow for uniform regulations within its 15 Member States (now 25) so that medical devices could be sold and distributed following one set of regulations. Landmark legislation and agreement was in the form of Medical Device Directive 93/42 EEC that allowed for the conformity, or the CE marking of medical devices. This harmonisation of device regulation in the EU began the process of 'global harmonisation' of medical device regulation that continues throughout the industrialised countries to this day.

The EU Medical Device Directive created a mechanism that established notified bodies, i.e. organisations that would be responsible for granting the conformity mark, the CE mark, to medical device manufacturers. In this Directive, it was established that the manufacturers needed to comply with ISO 9001 quality system regulations and ISO DIS 13485 quality systems for medical devices.

23.5   What is the CE mark?

The CE mark is the EU conformity marking that provides a basis for ensuring that a product will conform to certain essential requirements. Most medical devices to be marketed in the EU must have a CE mark, although Class 1 devices do not necessarily need this marking. There are always exceptions to the rule.

The essential requirements for a CE mark as stated in the Medical Device Directive 93/42 are 40 items written into this directive that cover all known sources of danger to the patient. The objective in establishing the essential requirements in development of these Directives was to consider all areas where a company might do something, or fail to do certain tasks, that could result in a risk to the patient. It was a risk assessment analysis that resulted in these 40 items.

These 40 essential requirements can be broken down into six major classifications:

1. Internal production control: that is, elements where a company must perform certain tasks to ensure control over their production.
2. The essential elements specify the type of examination that a notified body can carry out to ensure that a company is manufacturing and designing medical devices according to the regulations.
3. It establishes how one can assess that a device conforms to its intended use and needs.
4. It lays out guidelines for quality assurance (QA) systems in production that must be followed.
5. It specifies how to verify that the product is doing what it was designed to do.
6. It specifies guidelines for how a company will pursue a company-wide quality assurance system.

These groupings make up those 40 elements that form the basis for the regulation within the EU.

Essential requirements for a CE mark
- Internal production control
- EU type examination

- Conformity of type
- Production quality assurance
- Product verification
- Full quality assurance

23.6   Differences between FDA and EU regulations

The FDA's primary role, as established by the Congress of the USA in regulating medical devices, is protecting the public health. Everything beyond that is secondary to the FDA's mission. In the EU system of regulation there is an emphasis on the importance of a standardised 'internal market' as well as protecting public health. Another difference between the USA and EU systems is that the FDA individually reviews every device that is submitted to it and determines whether that device may be marketed. A company may not put a device on the market in the USA until, in some manner, they have notified the FDA.

For Class 2 and Class 3 devices, FDA must approve the device before the company is allowed to sell the device. By contrast, in the EU system, the company submits its data and information to the notified body, which is a private organisation chartered through the EU. That notified body has the ability then to grant or issue the conformity mark – the CE mark – and the company is allowed to put its medical device into the market place. Under this scheme individual governments do not review the decision of the notified body. In the EU scheme, the notified body is given the authority to clear, or to allow, the medical devices to be sold.

Differences between EU and USA device regulations
- FDA individually reviews every device submitted and determines marketability.
- No government in the EU reviews the notified body decisions on granting the CE Mark.
- FDA relies heavily on Federal Regulations.
- EU relies on voluntary standards to ensure safety and efficacy of devices.

The USA FDA relies heavily on the US Congress and Federal Regulations as its guide and its rule-making body to determine how the FDA will proceed in allowing medical devices on the market. The EU relies heavily on consensus-built voluntary standards, such as the three medical device directives and the voluntary standards of ISO, to ensure the safety and the efficacy of medical devices. These differences are striking when it comes to putting a new device into the market place, as shown in the CD lecture for Chapter 23.

In the EU, the 'competent authority' is the agency in that particular country responsible for the control of medical devices.

For example, in the UK the competent authority is the Medical Devices Agency. A company that is developing a medical device – such as a bioceramic bone graft material – only needs the CE mark that it obtains from a notified body within the EU in order to sell the device. However, the Medical Devices Agency in the UK or the competent authority in any country has the right and the mission to oversee and ensure that devices are performing as stated. So, a competent authority is able to prevent a device from being sold if there is a problem. It is able to take other actions necessary against medical device manufacturers to ensure the devices are being sold and maintained within the framework of the EU conformity marking system.

23.7   How do companies get through the FDA process?

The reader is requested to go to the Chapter 23 lecture on the CD for a discussion of the many steps involved in obtaining FDA clearance for a new medical device.

23.8   Summary

The FDA provides regulatory guidance and approval for biomaterials and devices in the USA. There are three classes of device: Class 1 devices are those with the lowest risk to the patient; Class 2 offer intermediate risk and Class 3 devices are those that are life threatening if they fail. Most implants, such as fracture fixation devices, are Class 2 whereas devices that supplement or replace the function of organs, such as artificial heart valves are Class 3. In Europe, biomaterials and medical devices are regulated by EU controls and must obtain a CE mark by satisfying the International Standards Organisation (ISO) standards of manufacturing and record keeping.

23.9   Reading list

Borozino J.D. (editor), *The Biomedical Engineering Handbook* (Chapter 193), Boca Raton, Florida, CRC Press, 1995.
Hench L.L. and Wilson J. (editors), *Introduction to Bioceramics* (Chapter 19, Appendix 2), Singapore, World Scientific, 1993.

Technology transfer

LARRY L. HENCH
Imperial College London, UK

24.1 Introduction

The increasing lifetime of patients makes it important to develop new biomaterials, improve device design and create new repair therapies that have improved survivabilities. All new products and therapies must be proven to be safe before they can be used in the clinic. Chapter 23 describes the regulatory process required to ensure product safety. This chapter describes the steps required to achieve regulatory approval and eventually a profitable product. This process is called *technology transfer*.

It is important to recognise that technology transfer involves a series of paths. Each path is followed by different personnel with different skills sets. Each path has a different output and must have a separate budget. Effective development of a new healthcare product requires completion of all of the paths. Profitability often depends on the time and cost invested in each path. Time and expense of the paths are additive and must be accurately projected and followed if the technology transfer process is to succeed.

24.2 Technology transfer paths

Table 24.1 summarises the five paths required for successful technology transfer. Details of the steps in each path are shown in Fig. 24.1. Each technology transfer path has a timeline, which is governed by the serial

Table 24.1 Technology transfer paths

	Path	Output
1.	Research	PhDs and publications
2.	Patent protection	Patents issued
3.	Market assessment	Potential profit and risk
4.	Technology demonstration	Prototypes, quality assurance, potential profit margin
5.	Production	Profits

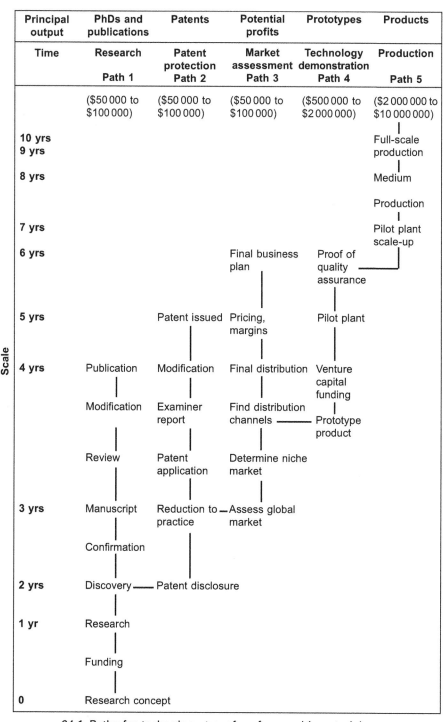

Principal output	PhDs and publications	Patents	Potential profits	Prototypes	Products
Time	Research	Patent protection	Market assessment	Technology demonstration	Production
	Path 1	Path 2	Path 3	Path 4	Path 5

24.1 Paths for technology transfer of a new biomaterial.

sequence of steps, indicated in Fig. 24.1. The output of each path is summarised in Table 24.1. The research path (path 1) is often much longer than that shown in Fig. 24.1, due to the need for cell-based or animal testing of new materials. The timeline indicated is a minimum, as is the projected cost of the research. Likewise, the timeline for patent protection can be substantially longer than shown if the invention is made in an area with many products already in the market place, which is typical for the biomaterials field. All five paths must be followed in order for a research concept to become a clinical product and commercial success, i.e. to become a product that can be produced and sold in the market place with a reasonable return on investment.

Although a specific length of time is indicated for the steps shown in Fig. 24.1, in reality there will be a distribution of times for each step that will be unique for a given technology, product and organisation.

The timelines shown in Fig. 24.1 tend to be towards the short end of the distribution and thus can be used as a good measure of effective technology transfer. If time increments are longer than those shown in Fig. 24.1 there is likely to be a problem somewhere in the organisation and the technology transfer process will be delayed.

In practice, the length of time for each of the five paths shown in Table 24.1 is nearly optimal. It is seldom possible to decrease the length of time required for the research, patent protection, market assessment or technology demonstration phases of a new product. This assumes the product involves substantially new technology. Incremental improvements in previously existing processing methods or products take perhaps a third less time in the technology demonstration phase because the pilot plant facilities are already in existence. Timelines for R&D and patent protection are nearly invariant, regardless of the level of innovation being pursued. It is seldom possible to shortcircuit any of the individual steps in Table 24.1 without suffering expensive delays later.

24.3   Efficient technology transfer

Figure 24.1 shows technology transfer as five parallel paths. It is optimal to pursue all five paths in parallel, as illustrated, rather than in sequence, for several reasons: (i) shorter cumulative time, (ii) feedback of information between paths, (iii) maintenance of momentum, and (iv) lower total cost.

If it is possible to pursue the five technology transfer paths in parallel, as shown in Fig. 24.1, then a cumulative time for a successful technology transfer process is approximately 8 years. However, if it is necessary to complete each path prior to commencing the next, the cumulative time is nearly doubled, to 14 years. Often, this is the case because the costs associated with patent protection and technology demonstration are usually considerably larger than research costs. New layers of management become involved in the decision-making process as one moves from path $1 \rightarrow 2 \rightarrow 3 \rightarrow 4$. The number of decision makers increases. Since the project costs of path $2 \rightarrow 5$ are substantially higher than usually budgeted in universities or small companies, the time required to evaluate the project, as well as the number of people required to evaluate, go up proportionally. The probability of approval goes down proportionally. A combination of factors often leads to an excessively lengthy serial technology transfer process. The major impedance step in the serial process is the transfer from path $3 \rightarrow 4$. The level of financial commitment goes up by a factor of 10 at this point. However, cost is not the only barrier in moving from path $3 \rightarrow 4$; additional personnel, management and facilities are equally important factors.

In order to demonstrate that the technology has production potential (path 4), it is essential to create a team composed of the scientist(s) who originated the discovery, engineers capable of scaling up the technology and experienced in designing the requisite equipment, along with technical staff, management, and financial planning and resources. The experience, skills, responsibilities and temperaments differ greatly among such a team. Consequently, there can be considerable time invested in achieving an acceptable schedule, plan of action, budget, and commitment to 'make it work' before path 4 can be activated.

The technology may not work in the demonstration path 4 without a number of trials. Feedback of information from the technology demonstration team to the research team and vice versa, i.e. paths $1 \leftrightarrows 4$, is essential. Often, creative scientific personnel will have moved on to other interests and are not enthused about returning to an 'old' project when problems arise. New personnel must be recruited and trained. The net effect is a lengthening of path 4. These difficulties can arise whether the technology transfer effort is occurring within the university or a company or with an 'arms length' licensing agreement between the university and a large corporation.

Usually, funds to pursue a Technology Demonstration project will not be approved without completion of a marketing and business study, path 3. The marketing and business analysis will attempt to project cost/benefit ratios, capital required, size of market, time to reach the market, percentage of market penetration, competitive position of the new technology, lead times over the competition, profit margins, effect on existing corporate products, etc. University science and engineering departments do not have the staff or experience to make this analysis. Usually, a licensing agreement must be in place to move from path 2 to 3. Delays often occur because funds are seldom available to initiate path 3 until patents are issued, the end point of path 2, and licence agreements are signed. This takes time and money.

The greater the advance of the new technology, the more difficult it is to make marketing and business assessments. Therefore, the greater the potential of the new technology, the greater is the risk and the longer is the time required to pass the judgement that it should be supported to enter path 4 and become a technology demonstration programme.

There are two primary difficulties in pursuing Fig. 24.1 paths 2, 3 and 4 entirely in parallel: (i) the 5× and 10× increases in cost of moving from prototypes to pilot plant scale operations, and (ii) the need to complete market and business assessments before larger budgets can be approved. Consequently, the staggered parallel paths (2 → 3 → 4) shown in Fig. 24.1 are usually required by economic realities.

The scale up from pilot plant to production (path 4 → 5) requires even more capital and market assessment. It is critical to know the size of the production facility required to achieve projected profit margins; however, production rates must be targeted towards sales projections. Thus, identification of profitable niche markets in the first years of scale-up is a key requirement for projecting the profit/risk ratio for the new technology. Large corporations have the expertise to make these assessments, but their large overheads inflate the required margins of profits to be a successful business.

Small start-up companies, conversely, have low overheads, but often lack the ability to assess accurately the multiple factors involved during the transition from path 4 → 5 and the steps in path 5.

24.4   Factors affecting rapid technology transfer

Rapid technology transfer depends upon several key factors:

1. The researchers responsible for creating the technology (path 1) and the patents (path 2) must participate in the early steps of path 3 (market assessment) and path 4 (technology demonstration). Funds to compensate them for the time committed should be incorporated within the budgets of paths 3 and 4 in order to ensure the researchers' co-operation.
2. Transitions between paths 1 → 2 → 3 must be quick and efficient. To achieve this end requires a management structure that reviews technology and patent disclosures rapidly. It is better to make a decision not to file a patent than to delay for many months. The research team must be made aware of the necessity of early disclosures of findings to begin the patent search while the reduction to practice is under way.
3. Licence agreements need to be flexible to allow rapid pursuit of paths 2, 3 and 4 simultaneously even though information to conduct the market assessment (path 3) is incomplete. There should be several iterations of the market assessment as data from path 4 becomes available.
4. Milestones with specific timelines need to be agreed upon by the research, management, marketing and product development teams responsible for paths 3 and 4. These timelines need to be co-ordinated with budgets and

reward incentives need to be built in to ensure the effort is made to meet the milestones. Management must keep in mind that university research teams have little, if any, motivation to participate in paths 2, 3 and 4 unless incentives are designed into the licence agreements. Future royalty agreements common to many licensing arrangements offer little incentive to researchers because they involve uncertain payoff schedules.

5.   Sufficient capital must be available to pursue paths 3 and 4. The capital should be available in stages that correspond to performance milestones for each path. Researchers may be reluctant to agree to milestones and specific timelines because it is counter to the research culture; however, tying budgets to performance is one way to ensure that the capital will be properly targeted towards the output of the paths instead of being used to continue research. One of the easiest ways to delay the technology transfer process is to continue to pursue research in path 1 with limited capital resources instead of moving the programme into paths 3 and 4. This often happens in universities because salary and promotion depend on publications, the primary output of path 1.

24.5   Alternative routes to commercialisation of biomaterials

When technology is created within a corporation, there is usually a management structure in place to make the decisions and budgets for paths 2, 3, 4 and 5. Milestones and timelines are often imposed as part of the job requirements of the teams involved. In contrast, when the technology is created within a university or a government laboratory, there is seldom a management structure or budget for paths 3 and 4. The approach to technology transfer is typically a licence agreement with a company. There are five alternatives for implementing the technology transfer process, as illustrated in Fig. 24.2.

Each option shown in Fig. 24.2 has a different degree of risk, time and personal capital investment associated with it. The height of the bar is proportional to these factors. The width of the bar represents potential pay-off; the greater the risk, the greater the potential pay-off.

Details of these alternatives and an assessment of the risk/reward ratios can be found in the book 'Sol–Gel Silica', by Hench, pages 133–150.

24.6   Summary

Five paths of technology are required to bring the concept of a new biomaterial or device to commercial production and clinical use. Path 1: research producing publications and PhDs; path 2: patent protection yielding patents; path 3: market assessment providing output and potential profits; path 4: technology

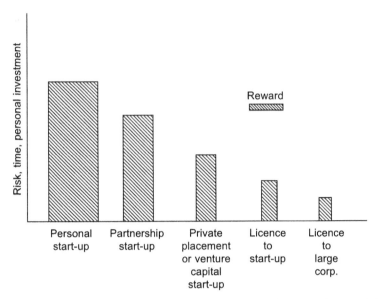

24.2 Alternatives to achieve commercialisation of a new biomaterial or tissue-engineered product.

demonstration yielding pilot plant production of prototypes; path 5: production of products under regulatory control. The cost and timelines required increase for each path. A typical technology transfer process in this field takes at least 8 years.

24.7   Reading list

Hench L.L., From concept to commerce: the challenge of technology transfer in materials, *MRS Bulletin XV*, Vol. 8, pages 49–53, 1990.
Hench L.L., *Sol–Gel Silica: Properties, Processing and Technology Transfer* (Chapter 11), Westwood, NJ, Noyes Publications, 1998.

25

Ethical issues

LARRY L. HENCH
Imperial College London, UK

25.1   Introduction

The objective of the science and technology described in Chapters 1–24 is to enhance the quality of life of an ageing population. We have seen that there are three alternative routes to the repair or replacement of diseased, damaged or worn-out tissues and organs:

1. transplants;
2. implants;
3. tissue engineering.

These options are summarised in Fig. 25.1 and discussed in detail in CD Chapter 25 (Ethical issues of implants) (CD Fig. 25.3).

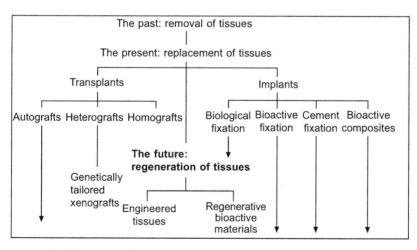

25.1 The revolutionary changes in the treatment of orthopaedic defects over the last century. (From L.L. Hench, 'The challenge of orthopaedic materials', *Current Orthopaedics* **14**, 7–15 (2000).) (CD Fig. 25.3).

There are uncertainties associated with each of the three optional routes to repair and replacement of tissues and organs. Autografts, transplants from the patient's own body to another location in the same patient, are considered the 'gold standard' but there is a limited amount of material available and morbidity is associated with the second site surgery. Heterografts, transplants from another human, are restricted due to a limited supply, cost and a requirement for lifetime immunosuppression drugs (CD Chapter 25). Xenografts, transplants from another species, require genetic modifications, as discussed in CD Chapter 25, or require chemical modifications to eliminate immune rejection.

Use of implants, prostheses or artificial organs to repair or replace body parts, described in Chapters 12–17, are used much more than transplants because they are more readily available and are seldom rejected by the body's immune system. However, it is important to remember that all implants and artificial organs are man-made. They have three major limitations: (1) no ability to self-repair, (2) no ability for self-adaptation, (3) their mechanical and biological properties are a compromise for the living tissues and organs being replaced. Thus, no man-made spare parts are as good as the living parts they replace. There will always be relative success and relative failure of implants, just as there is for transplants. There is a finite probability that an implant or transplant will fail and need to be replaced during the lifetime of the recipient. Engineering of tissues and organs, Fig. 25.1 (CD 25.3), provides a third option to circumvent the limitations of transplants and implants, as described in Chapters 18–22. The objective is to use the patient's own cells or an immunotolerant 'universal' cell source to grow replacement tissues or organs *in vitro* (tissue engineering) followed by transplantation into the patient. Alternatively, tissue-engineered cells, with or without genetic modification, can be placed into a patient and stimulate regeneration of tissues and organs (Chapter 18).

25.2   The ethical problem

All three options for repairing and replacing tissues and organs summarised above (Fig. 25.1) have limitations and uncertainties. All three offer a compromise. A decision has to be made as to which option to use for a patient. That decision must be made by the physician, the patient and the family, often within the constraints of limited resources. The problem is that ethical dilemmas arise from a conflict of uncertainties. Making judgements of the relative value of alternative medical treatments has become more and more difficult. The difficulty is the expansion of alternatives. Before implants and transplants became available, the decision for the dentist or surgeon and the patient was simple. If a tooth hurt, it came out. If a hip hurt, you quit using it. If a heart stopped, you died. One problem, one solution.

Today's technology-based medicine offers many more alternatives, not all of equal value in terms of risk, benefit and cost. Consequently, these alternatives often produce uncertainties and ethical and moral dilemmas. False expectations of 'miracle' cures and unwillingness to accept the finite probability of failure of implants and transplants add to the ethical problem. A failed prostheses is very bad, for many reasons. The person is older and recovery from major surgery is more difficult. The host tissues are less capable of repair. There is always some chance the patient will not wake up from the anaesthesia, die from a blood clot or heart failure, or get infection during or after surgery. These complications are all bad. They are also all real and finite. For every implant or transplant that is a cure, there is a certain probability that some other implant or transplant will be a 'failure'. For example, as discussed in Chapter 13, total hip prostheses have been an enormous success for millions of patients. However, there is a finite probability of failure that requires revision surgery; it ranges from about 3 to 5 in 100 cases for the first 3 to 5 years after an implant to as much as 10 to 15 per 100 after 10 to 15 years. The cumulative effect of these failures is large, e.g. of approximately 40 000 hip implantations per year in the UK more than 5000 will be revision surgeries. Similar statistics exist for all implants and transplants and lead to uncertainties as to who should make the decisions on relative priorities.

25.3   Moral uncertainties

In the UK, medical decisions greatly depend on National Health Service (NHS) resources. Long waiting lists for elective total hip replacements have resulted. Several years ago Professor Bill Bonfield, previous Director of the IRC in Biomedical Materials, at Queen Mary College, University of London, showed me a broken femoral stem of a total hip prosthesis. He said, 'The patient is 90 years old and this is his fifth failed implant. Should he be implanted with a sixth?' This is an ethical dilemma for the surgeon, patient and the hospital when there are thousands of younger patients still waiting for their first implant. These socio-economic issues of healthcare distribution are discussed in the book *Science, Faith and Ethics* in the reading list.

Professional ethicists and moral philosophers recognise that there are serious problems in resolving the type of moral uncertainties described above. In the introductory chapter of *Contemporary Issues in Bioethics*, Beauchamp and Walters write, *'Some moral disagreements may not be resolvable by any of the means discussed. We need not claim that moral disagreements can always be resolved, or even that every rational person must accept the same method for approaching such problems. There is always the possibility of ultimate disagreement.'*

This admission of the limitations of ethical theories by professional ethicists poses a problem for solving ethical problems associated with use of implants, transplants and tissue engineering. The writings of John Stuart Mill and his successors defend an approach called *utilitarianism*. Their position, that an action is right if it leads to the greatest possible good consequences or least possible bad consequences, is reasonable to most people. Moral rules, thus, are the means of fulfilling individual needs and also of achieving broad social goals. A moral life is measured in terms of values and the means of producing the values. Doing 'good' benefits both yourself and society.

However, the eminent German philosopher Immanuel Kant and his successors, such as W.D. Ross, argue that moral standards exist independently of utilitarian ends. Their ethical theory is termed '*deontological*', from the Greek word '*deon*', which means '*binding obligation*'. Thus, deontologists argue that a moral life should *not* be conceived in terms of means and ends. An act is right not because it is useful, but because it satisfies the demands of some overriding principle of obligation. They maintain that human decisions should be weighed against a higher authority, such as the Revealed Word, the Bible, the Torah or the Koran. Kant stipulates, '*One must act to treat every person as an end and never as a means only*'.

The differences in ethical theories are fundamental and are difficult, if not impossible, to resolve. Moral philosophers who conclude that 'right is relative' maintain that the concept of right and wrong and good and evil has evolved from man's social interactions. Thus, what is right, including medical treatments, depends upon the consequences of action in a specific social context. In contrast, moral philosophers, such as Kant, who conclude that 'right is revealed' believe that the concept of good is innate to the individual. Action should be based on the revealed perception of what is right or wrong and therefore is independent of its social context. This leads to theoretical moral disagreements, referred to by Beauchamp and Walters, which may never be resolved. The conflict between these theories leads to ethical dilemmas. The consequences of these moral uncertainties in the use of implants, transplants, healthcare distribution, use of embryonic stem cells and genetic alteration of life, as well as birth and death control, are the subject of the book *Science, Faith and Ethics*, in the reading list.

Two practical ethical dilemmas that need to be resolved illustrate these ethical dilemmas. (1) Our technology-based society has generated infinite desires in people to live a long life. However, there are only finite resources to maintain our quality of life. This imbalance between infinite desires and finite resources creates severe ethical dilemmas. (2) Everyone desires a long life. The dilemma is how to achieve a balance between *quantity* of life and *quality* of life.

Many situations in the use of biomaterials, artificial organs and tissue engineering are uncertain in their moral dimensions. Several examples from my experiences with biomedical materials illustrate the problem. One individual who was president of a company with rights to commercialise our Bioglass®

implant technology raised many millions of dollars, and brought two products to the market, benefiting thousands. He then stole several millions of dollars from the company and went to a federal penitentiary. Another individual in a different company was also given legal rights to commercialise Bioglass® implant technology. He did not bring any products to the market, and thereby thousands of patients did not benefit. He succeeded in corporate life in spite of this failure. A third individual developed a similar bioactive material, used it for implants, sold his company for several million dollars, and several years later many of the implants failed. A fourth person put a medical material on the market. It succeeded in three clinical applications. However, in a fourth application the material degraded and caused great pain in thousands of patients, leading to corporate bankruptcy and enormous suffering.

Of these four individuals, who was the most morally upright? We do not have the basis to judge. The uncertainties are too great. The courts are given the right to judge. But we, as individuals, do have the right and the responsibility to learn from these examples and the consequences of their failures.

25.4   General moral principles

There is agreement among moral philosophers regarding three general principles that can be used to guide behaviour and make ethical decisions. The moral principles are respect for autonomy, beneficence and justice.

The *principle of respect for autonomy* refers to the concept of personal self-governance. It is the opposite of slavery. It assumes that individuals possess an intrinsic value and have the right to determine their own destiny. This principle of a person's right to choose is the fundamental issue that governs most medical situations. Many moral philosophers consider this principle to rank highest in any hierarchy of ethical principles.

The *principle of beneficence* is that an action or decision should not inflict harm on another, should prevent or remove harm, or promote good to another. This principle says that it is morally right to aid, or to prevent harm to, another. Conversely, it is morally wrong to harm, or to prevent aid to, another.

The *moral principle of justice* requires that like cases be treated alike. This simple concept is called the 'formal principle of justice or equality'. However, it does not specify *how* to determine equality or proportion in making moral judgements. Therefore, this principle in its formal sense provides little guidance for decisions regarding conduct or behaviour concerning alternative means of healthcare.

25.5   Principles of distributive justice

To solve the problem outlined in the preceding section, moral philosophers have developed a variety of alternatives called *material or distributive principles of justice*. Their goal is to provide a basis for judging the *relative* needs of people. This is necessary and important because, as we are well aware, people and their needs are not equal, especially with regard to health and health care needs. For example, some people are healthy most of their lives. Some individuals experience a lifetime of illness or are even born with genetic defects. Their needs are substantially different, so the decision as to what is a just distribution of healthcare is not straightforward. Since people are not equal, how can they be treated alike, as the formal principle of justice requires?

A variety of principles of distributive justice have been proposed and discussed in books given in the Reading list. They include ten alternatives listed in Fig. 25.2.

Principles of distributive justice

Alternatives

1. To each person an equal share
2. To each person according to individual need
3. To each person according to acquisition in a free market
4. To each person according to individual effort
5. To each person according to societal contribution
6. To each person according to merit
7. To each person according to age
8. To each person according to status (nobility)
9. To each person according to gender
10. To each person according to race

25.2 Principles of distributive justice.

There are large differences between these alternatives. People will often vigorously defend one of them as just and others as unjust. Consequently, most societies use several of the above principles in combination, in the belief that different rules apply to different situations. The NHS in the UK was created to administer healthcare according to the principles of equal share or equal need. This ideal has become a fiscal impossibility. The growing failure of the NHS to meet the near-infinite demands on it has led to growing criticism. CD Chapter 25 discusses these issues.

Many of the difficulties in establishing a national healthcare policy in the USA, which are discussed in Chapter 8 of *Science, Faith and Ethics*, are due to the enormous disparity in the principles of justice listed above. Advantages and disadvantages of the alternatives for distributing care that have life and death implications are discussed Chapters 7, 9 and 10 of *Science, Faith and Ethics*. Ethical conflicts abound. Some of the problems involved in the widespread use of implants, as described above, are due to the fact they are so readily available. Except for cost, there is little moral concern about the use or distribution of implants. If you want one, and can afford it, you can usually have it. The same cannot be said for transplants. Many additional ethical issues are involved, as discussed in CD Chapter 25. Let us summarise some sources of ethical conflict that are created by the difficulty in resolving conflicts between the three great principles of ethics.

25.6   Consequences of the theoretical problem

Beauchamp and Walters summarise the biggest problem facing ethicists and moral philosophers at the present time: '*The problem of how to value or weigh different moral principles remains unresolved in contemporary moral theory.*' This means that, when the *principle of respect for autonomy* is in conflict with either the *principle of beneficence* or the *principle of justice*, there is no acceptable means of resolving the conflict.

The consequence of this theoretical problem is that it leads to uncertainty in assessing an ethical response in individual cases. Guidelines for ethical behaviour can be developed for large population groups, but they may not be accepted by individuals within the group. Individuals often consider general guidelines or restrictions to be unjust if they are excluded. Thus, uncertainty in the relative importance of the three ethical principles leads to conflict between individuals and between individuals and the group.

For example, consider the situation where an implant fails. Conflict may arise between the patient and the surgeon, hospital or manufacturer, or all three. Why is there conflict? The patient chose to have the implant, risks were reviewed and informed consent was obtained; thus the patient's autonomy was respected. The individual case history indicated to the surgeon that the implant and procedure selected had a high probability of success, thereby fulfilling the principle of beneficence. Conflict results, however, when the patient perceives that the principle of justice has been violated. The patient expects not only equal treatment, but also equal results. The patient and his/her family do not care about statistics, and that 85 or 95% of similar cases treated the same way succeeded. They care only that their case has failed.

25.7   Sources of conflict

Figure 25.3 summarises several sources of ethical conflict. One of the most important in the medical field can be stated as: '*The conflict is due to an unjustified expectation of equal consequences of an act instead of equal performance of the act*'.

The principle of justice specifies only that '*like cases be treated alike*'. However, because individuals are different, the results can be different even if the treatment is the same. The difference in results versus expectations can

---

Sources of ethical conflict

- Perception that principle of justice has been violated.
- Unjustified expectations of equal consequences of an act instead of equal performance of an act.
- Emphasis that technology provides certainty.
- Failure to understand that all technological solutions have some risks. A risk-free life is impossible.
- Greed feeds on technology and conflict.
- Legal system that fails to recognise that all individuals are different.

25.3 Several sources of ethical conflict.

be perceived, wrongly, as unjust. What are the reasons for unjustified expectations of implant success? Three factors, at least, are involved: *human nature, technology* and *greed*.

It is *human nature* to want the same things as others. This expectation feeds our market-driven economy. The same is true for implants and transplants. People do not desire to live with pain, as they become older. This is reasonable. They learn from the media, their physician or friends that certain implants eliminate pain, and therefore they want an implant if they have painful joints. They do not hear, or are unwilling to accept, that there is a finite risk associated with the surgery and a finite possibility of the failure of an implant. It is human nature to hear only what you want to hear. This results in unjustified expectations and a conclusion of receiving unjust treatment if difficulties arise.

Technology amplifies the problem. New developments in implants are promoted as superior even when long-term data for large populations of patients are not available. We live in a technological age where most people want and expect the latest, be it electronics, cars or implants. Along with the latest comes the expectation that the latest is the best. This is often unjustified, but the perception still exists. Rapid changes in technology also lead to a proliferation of choices. The surgeon and the patient no longer are limited to one decision: Should a hip joint be replaced with a prosthesis? Instead, a series of decisions must be made with regard to: type of stem, type of cup, type of fixation, etc. The statistical basis for risk assessment and beneficence becomes progressively more uncertain, the greater the options. The patient may well equate more options with a greater chance of success. This is often false. In fact, the reverse may be the case, i.e. success in a large population decreases as the number of options increases.

Greed can feed on the above factors. As more people want and receive implants, the potential for profits increases proportionally. As more options become available, it is more likely that products will be promoted for the sake of novelty and image rather than for statistically based improvements in beneficence over long times of use. Economic pressures build to introduce new implant products with only minimal standards of testing in order to have something new to offer. Tests that show that problems may occur are undesirable in this context and therefore are avoided unless required by regulatory pressures. Research to obtain solutions to long-term reliability problems is often not done, because to do so is to admit that long-term reliability is a problem. Thus, the implant field grows in volume but not necessarily proportionally in beneficence to the larger number of patients. One consequence is an ever-increasing escalation in healthcare costs and expanding numbers of revision surgeries.

25.8   Specific ethical concerns about biomaterials

There is a great need to promote testing to avoid long-term complications and implant revisions. We need to achieve >85–95% success for implants over 10–20 years. Failure analysis needs to be done for all implants in use or proposed for use, in order to provide a statistical basis for establishing beneficence for the patient. Figure 25.4 (CD Fig. 25.10) illustrates the type of analysis needed. Clinical results can be classified as those that result in high beneficence (top curves) or low beneficence (bottom curves). Moderate beneficence lies in-between. The ethical *principle of beneficence* requires that an implant meet the high standard of the upper curve because otherwise the principle 'first do no harm' may be violated. In other words, any implant that performs in limited trials for 3–7 years as indicated in the lower curve, should not be put into general use. Any implant that is in general use with results similar to those in the lower curve should be removed from use.

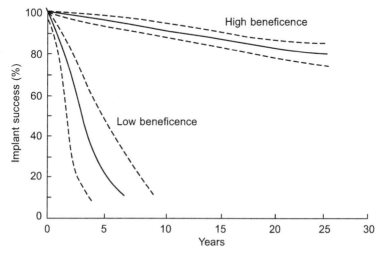

25.4 Comparison of implant failures as a function of time for high vs. low beneficence to the patients (CD Fig. 25.10).

Implants of moderate beneficence should be subjected to regulatory monitoring to determine the reasons for failures and research funded to improve the performance until the upper curve behaviour is achieved.

Figure 25.4 (CD 25.10) also illustrates that there exists a distribution of results that broadens with time. Statistical data that can be used to generate such curves need to be compiled by the professional societies for all the prostheses now in clinical use. Patients need to be informed of their expected benefit with respect to this distribution of results. This is one of the few ways to counter false expectations of results.

25.9   Summary

Many researchers, clinicians and manufacturers of implants have been exposed to the consequences of ethical conflicts, which can arise when the *Principles of beneficence and justice* are not reconciled. The expectations of the population with respect to implant success will continue to rise. Thus, implants must have increased long-term reliability. Failure to ensure this will result in negative consequences for individual patients. Failure will also produce increased governmental regulations and controls. These controls will increase development costs and produce a negative spiral in which fewer manufacturers will be able to afford to produce fewer products and will develop fewer new materials and applications. This negative scenario can be avoided by a concerted effort to improve the long-term performance of all types of implants. The overriding principle '*Respect for autonomy*' must be of concern to all of us. An individual depends on information from all sources to make crucial decisions. Such information should always be the best and most complete and be unsullied by commercial or personal preferences. Only by these means can the best decision be made for specific clinical problems. The concerned reader should pursue this topic in the CD Chapter 25 lecture and in the Reading list.

25.10   Reading list

Beauchamp T.L. and Walters L., *Contemporary Issues in Bioethics*, 3rd edition, California, Wadsworth Publishing Co., 1989.

The British Medical Association, *The BMA Guide to Living with Risk*, New York, Penguin Books, 1990.

Callahan D., *What Kind of Life: The Limits of Medical Progress*, New York, Simon and Schuster, 1990.

Cauwels J.M., *The Body Shop: Bionic Revolutions in Medicine*, St Louis, C.V. Mosby, 1986.

Hench L.L., *Science, Faith and Ethics*, London, Imperial College Press, 2000.

Hench L.L. and Wilson J. (eds), *Clinical Performance of Skeletal Prostheses*, London, Chapman & Hall, 1993.

Hilton B., *First Do No Harm: Wrestling with the New Medicine's Life and Death Dilemmas*, Tennessee, Abingdon Press, 1991.

Index

Printed and bound by CPI Group (UK) Ltd, Croydon, CR0 4YY

03/10/2024

01040435-0012